SPRINGER
LAB MANUAL

B. Grout (Ed.)

Genetic Preservation of Plant Cells in Vitro

With 17 Figures and 15 Tables

Springer

Dr. BRIAN GROUT

Crisp Innovar Ltd.
Glebe House
Station Road
Reepham
Norfolk NR 110 4NB
United Kingdom

ISBN 978-3-642-48968-6 ISBN 978-3-642-78661-7 (eBook)
DOI 10.1007/978-3-642-78661-7

Library of Congress Cataloging-in Publication Data. Genetic preservation of plant cells in vitro / B. Grout (ed.). p. cm. — (Springer lab manual) Includes bibliographical references and index.
1. Germplasm resources, Plant—Storage—In-vitro. 2. Germplasm resources, Plant—Cryoperservation. I. Grout, B.W.W. II. Series. QK981.7.G46 1995 581′.0724—dc20 94-47310

Cover design: Struve & Partner, Heidelberg
Typesetting: Best-set Typesetter Ltd., Hong Kong
SPIN 10037334 39/3130-5 4 3 2 1 0 – Printed on acid-free paper

Foreword

The long-term storage and maintenance of viable plant cells and organs is an area of active concern across the range of pure and applied plant sciences. In academic, government and commercial laboratories, the extended storage of propagules of one sort or another, with maximum protection of the genome from mutation and altered expression, is often a very necessary activity that can draw heavily on resources and effort.

However, preservation per se is typically not an activity in its own right, but a facilitating technology that is part of a larger programme of work. Consequently, there are many laboratories that do not have the benefit of a specialist in storage technology, and have to delegate the responsibility to individuals, or teams, who are faced with a daunting learning curve. To maximise the chances of success, in the shortest possible time and with minimum losses, these researchers need sources of reference that are authoritative and soundly based in practical experience.

It is with such a viewpoint in mind that this volume has been compiled, drawing upon the everyday, practical experience of scientists who are recognised for their pioneering involvement in storage technology as applied to plant tissues. Progress in the field has been rapid in recent years but an understanding of the basic processes is still a very long way from being complete, or even satisfactory. There are many questions still unanswered and just as many areas of investigation as yet untouched. It is likely that the readers of this volume will have a need, and a curiosity, that will draw them into these new areas to a greater or lesser extent, and it has been the contributors intention to give as much advice as possible to help in such circumstances.

In addition to the information we have provided, and the pointers we have given towards modifying procedures, the authors offer their best wishes for good luck and good results to those who take on the challenge of successful storage of viable plant cells and tissues.

Brian Grout

December

Contents

Contributors

ERICA E. BENSON
Department of Molecular and Life Sciences
University of Abertay Dundee
Bell Street
Dundee DD1 1HG
Scotland

JOHN G. DAY
Culture Collection of Algae and Protozoa
Institute of Freshwater Ecology
The Windermere Laboratory
Far Sawrey
Ambleside
Cumbria LA22 OLP
UK

BRIAN GROUT
Crisp Innovar Ltd.
Glebe House
Station Road
Reepham
Norfolk NR10 4NB
UK

KEITH HARDING
Department of Crop Genetics
Scottish Crop Research Institute
Invergowrie
Dundee DD2 5DA
Scotland

MARK R. MCLELLAN
Acer Environmental
Acer House

Howard Court
Manor Park
Nr. Daresbury
Cheshire WA7 15J
UK

Andrew V. Roberts
Division of Genetics
University of East London
Romford Road
London E15 4LZ
UK

Leigh E. Towill
USDA/ARS National Seed Storage Laboratory
Colorado State University
Fort Collins
Colorado 80523
USA

1 Introduction to the in Vitro Preservation of Plant Cells, Tissues and Organs

B.W.W. Grout

As a consequence of the developing technology for sustaining growth and development of plant materials in vitro, tissues from a great diversity of species are cultured in this way for an equally diverse number of reasons (Bhojwani 1990; Nijkamp et al. 1990; Pollard and Walker 1990). These will include transient structures such as isolated protoplasts and excised shoot and root meristems, that must proceed with development as a necessary adjunct to their survival, and somewhat more stable structures such as single cells, cell clusters, somatic embryos and fully differentiated plantlets. These may be maintained in liquid growth medium or on the surface of, or embedded in, medium made semi-solid with various gelling agents. Such in vitro cultures are employed in all aspects of basic and applied research in the plant sciences, encompassing molecular biology, biotechnology, crop improvement, plant production and basic metabolic and anatomical studies.

The successful management of in vitro plant systems must take into account the rate of growth of the material and the depletion of essential nutrients from the growth medium. Each culture vessel must be attended to as the contained material is about to outgrow the container and/or if growth is constrained by the lack of nutrient. At this subculture stage the material is removed from the original container, subdivided in some way if this is appropriate, and transferred to a fresh container with new growth medium. Despite its modern origins, in vitro plant technology is extremely labour-intensive, as the interval between each subculture or other handling of the culture vessel is likely to be in a range from 7 to 35 days. This creates a high, continual demand for skilled labour, media, culture vessels and space in controlled-environment growth chambers. In any facility, whatever its size, it is likely that there will be a significant proportion of material that is being maintained in vitro for some future purpose and not as part of any current multiplication or experimental programme. Consequently, there would be significant logistic and resource benefits to be gained by employing a storage system for these in vitro materials, that significantly extends the intervals between subculture and other handling operations. Use of such a storage system would also allow greater control of production schedules for in vitro material. For example, multiplying plantlets at a particular developmental stage might be required for commercial purposes and could be stockpiled as required, or appropriate starter cultures for large-scale fermentations might be stored for use as required. Prolonged maintenance of in vitro material also provides an effective system for establishing both active and base level germplasm collections, particularly for the conservation for the maintenance of biodiversity of those plant species that cannot be stored as true seed and are amenable to micropropagation technology (Withers 1987; Grout 1990a; Dodds 1991; Fay 1994).

The properties required of a successful storage system are the ability to:

- Minimise growth and development in vitro of the plant subjects to enable intervals between subculture and other handling to be significantly extended.
- Maintain the viability of the stored material at the highest possible level, together with the minimum risk to genetic stability.
- Maintain the full developmental and functional potential of the stored material when it is returned to physiological temperatures.
- Make significant savings in labour input, materials and commitment of specialised growing facilities.

It should be noted that a high level of genetic stability is inherent in plant materials growing in vitro at the more complex levels of structural differentiation, as a function of plant meristem structure and the integrated organisation of a plantlet, meristem or somatic embryo. When plant material is maintained in vitro at lower levels of differentiation such as isolated cell suspensions and callus masses, then the risk of spontaneous genetic alteration (somaclonal variation) is comparatively high (Scowcroft 1984; Karp 1989), with genetic changes that may include point mutations, amplifications or deletions of gene sequences and losses of partial, or whole, chromosomes (Karp and Bright 1985; Larkin 1987; Philips et al. 1990). As this risk diminishes as the level of organisation of the cultured material increases, and where choice exists, the preferred material to be maintained in vitro should be either a differentiated plantlet or somatic embryos capable of continuing their development (Ammirato 1983; Terzi and Loschiavo 1990; Priel and Beck 1991).

It is also important when storing materials that species and variety details, the origins of the particular cultures and their in vitro history up to the point of entry into the storage system are well documented for future reference. The storage containers themselves must also be unambiguously and indelibly marked and a representative sample encompassing the genetic diversity of the original plant cell/tissue population taken for storage. For protection, duplicate samples of each accession in the stored collection should be stored in separate storage containers or environments to minimise the risk of loss under unforeseen circumstances. Given that there is a risk of non-lethal injury and genetic alteration associated with any storage technique, particularly where non-optimal techniques are being employed, it is important that recovered cultures can be adequately compared to an agreed control standard. The material for storage should be appropriately characterised using criteria such as morphological characters, growth rates, aspects of metabolism and secondary production and a range of assays at the molecular level (see Harding and Benson, Chap. 8, this Vol.).

This volume is intended to provide a practical guide to the most widely used effective strategies for the storage of plant cell, tissue and organ cultures in vitro. In many instances, the techniques that are described are not universal, at least in terms of their detail, and the need for an element of empirical investigation must be accepted to develop the optimal storage protocol in a particular circumstance. To guide such an investigation, the concepts that underlie the various storage techniques, together with key citations from the appropriate literature, are presented in this introduction and in the more specific chapters following. The strategies that are included have been successfully applied to higher plant and algal materials and deal with situations where growth and development are minimised, but not halted, and those where growth is completely halted by freezing. The protocols are limited to those that have generated a body of literature reporting their successful application, and so offer the prospect of being successful in typical, practical circumstances.

1.1 Storage Techniques

1.1.1 Modification of the Gaseous Environment

It has been observed that growth rates of in vitro plant cultures are influenced by the volume and composition of the gaseous atmosphere inside the culture vessel (De Proff et al. 1985; Gould and Murashige 1985; Belcher et al. 1987), particularly with regard to compounds such as carbon dioxide and ethylene. However, it is evident that maintaining and controlling the individual gas atmospheres of large numbers of individual culture vessels would be complex, time-consuming and expensive, and so unlikely to provide cost-effective in vitro storage. However, the improving development of gas-permeable culture vessels (Reed 1991) opens up the possibility of restricting growth by controlling the composition of the bulk gaseous atmosphere of the entire culture room, which is much more practical than dealing with individual culture vessels, and may be worthy of further investigation.

Simple control of growth of both callus and plantlet cultures has also been achieved by judicious alteration of the diffusion distance between the plant tissues and the point of gas exchange between the culture vessel and the external environment. In a study with carrot plantlets differentiating from the surface of a callus mass, a range of culture vessels were constructed with identical gas headspaces, but different diffusion distances (Bateson et al. 1987). Increasing the diffusion distance was used to halve the rate of biomass increase and effectively double the period of subculture for 30 to 60 days. Whilst such a technique is simple, and has obvious potential, it has not been subsequently developed in a systematic way. Consequently, it must still be viewed as very much an experimental technique and will not be dealt with further in this volume.

1.1.2 Minimal Growth at Reduced Temperature

The most obvious and simple way of influencing the rate of growth of plant materials in vitro is by reducing the growth temperature of cultures from the optimum. This is particularly suitable for differentiated plantlets, and entire cultures, complete with growth medium and in the culture vessel, are used. Where the temperature can be reduced for prolonged periods without significant injury to the plant tissues, a basic storage technique can be developed (Withers 1985, 1987; Dodds 1991). In conjunction with reduced temperature, growth might be further limited by the addition of inhibitory levels of growth regulators or osmotic agents (Withers 1985). However, it has to be recognised that minimal growth conditions will vary significantly with species, variety and often with specific lines of plants, which means that careful tailoring of conditions has to be carried out to devise the most effective storage system.

1.1.3 Cryogenic Storage

The greatest stability of in vitro plant materials, with practical storage periods measured in decades, can be achieved by cryogenic storage at ultra-low temperatures, with liquid nitrogen being the most widely used, relatively non-hazardous cryogen (at −196 °C). Maintenance under these conditions effectively halts biological growth and development (Franks 1985; Grout and Morris 1987; Grout 1990a,b), for below −140 °C the rates of chemical and biophysical reaction will be too slow to affect cell survival. Consequently, material that can be brought to the ultra-low temperature and recovered from it without acquiring lethal injury, may be stored for extremely long periods. There is a small risk of injury to macromolecules, including the genetic material, in cryopreserved subjects, resulting from free radical activity during processing and from external, ionising radiation during storage. Any such injury will be preserved in the stored tissues and accumulate, due to the non-functioning of repair mechanisms at the cryogen temperature. Comfortingly, projections based on studies of animal cells in frozen storage suggest that the risk of radiation injury is insignificant when measured over at least several decades, if not centuries (Ashwood-Smith and Grant 1977; Grout and Morris 1987). In practice, cryopreservation provides a genuinely long-term storage technology but, as with all storage systems, there is a finite point, probably measured in decades, at which conserved material should be recovered and regrown in vivo before being re-evaluated and re-entered into storage. The challenge is to devise a protocol that allows in vitro plant material to be recovered from the cryogen at high viability, and structurally and functionally unaltered (Kartha 1985, 1987; Withers 1985, 1987; Grout and Morris 1987; Grout 1990a,b).

1.2 Principles of Cryogenic Storage

Techniques of cryopreservation are still relatively new for plant tissues, and a great deal of empirical work is likely to have to be done when attempting to cryopreserve a new, or recalcitrant, subject. Planning and execution of this work requires an understanding of the basic mechanisms underlying the cryopreservative procedures. Successful cryopreservation of in vitro plant cultures can be achieved by using either slow cooling or vitrification techniques, which differ in the way that the plant material is prepared prior to immersion in liquid nitrogen, and in their dependence upon a precisely controlled rate of cooling. Preservation of viability depends upon the ability to minimise the stresses of cryopreservation, described below, and protect against their damaging consequences. The goal for successful cryopreservation is to maintain a level of integrated structure and function compatible with high viability and normal activity upon restoration to physiological temperatures. The maintenance of viability cannot be the sole criterion of successful cryopreservation, as unaltered function at a precise qualitative and quantitative level is necessary under many circumstances (Benson 1990; Benson and Hamill 1991; Potter and Jones 1991).

1.2.1 Slow Cooling

The stages typically involved in cryopreservation by slow cooling are:

- Cold acclimation of donor material where possible.
- Pretreatment of explants with cryoprotectants and/or osmotic agents.
- Incubation in a cryoprotectant solution, in which the explants will be frozen and stored.
- Slow cooling, sometimes in a step-wise fashion, at a precisely defined rate to an intermediate temperature between 230 and 240°C. This provides a controlled amount of cryodehydration (see below).
- An optional, extended incubation at the intermediate temperature to continue cryodehydration, prior to direct immersion of the samples in cryogen.
- Thawing by immersion of sample containers in water at 35–40°C.
- Removal of cryoprotectants and restoration of osmolarity of culture medium to physiological levels.

A typical thermal history is shown in Fig. 1. The potentially lethal stresses imposed upon the biological material during cryopreservation result largely from osmotic effects that accompany the phase transition of water from a liquid to a more ordered state (Franks 1985; Taylor 1987). Reduced temperature per se will also impose stresses on temperature-sensitive species, as in chilling injury (Lyons et al. 1979; Wilson 1987) or denaturation of active macromolecules (Franks 1985).

Following cooling below its melting point, the incubating solution within the sample vessel that contains the material to be preserved will eventually freeze. Typically, there will be an amount of undercooling below this temperature before ice formation is initiated, or "seeded" by an appropriate nucleator (Meryman and Williams 1985). A nucleus, other than ice itself, is an ordered group of molecules sufficiently similar in their spatial relationships to the crystal lattice of ice that diffusing water molecules in their proximity will tend to order into genuine ice crystals. Water molecules themselves can form transient nucleii and ice crystals form from them by a process known as homogeneous nucleation (Franks 1985). Other surfaces can also act as nucleii, including dust, scratches on containing vessels and a range of organic and inorganic compounds. Ice formation from such nucleii is known as heterogeneous nucleation. The kinetic energy lost by the water

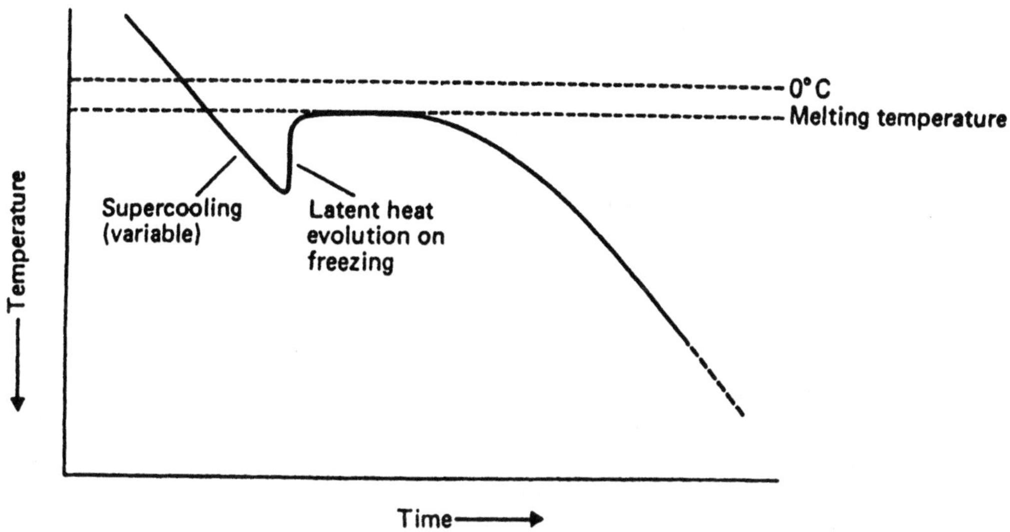

Fig. 1. A schematic thermal history for the incubating solution surrounding a sample of plant tissue during cryopreservation. The extent of undercooling and the exotherm following ice nucleation are indicated. (Ashwood-Smith and Farrant 1980)

molecules as they assume their crystalline conformation is released as an exotherm of latent heat which has to be dissipated before controlled cooling can be resumed (Meryman and Williams 1985).

Where a relatively dilute aqueous solution surrounding the biological subject is being cooled slowly, the extracellular medium can be viewed as a single, large compartment when compared to the cellular marterials of the plant tissues. The probability of ice nucleation in these circumstances is related to compartment size (Hobbs 1974), and these extracellular solutions, rather than those within the cells, freeze as a first event during cooling. Ice propagates through the extracellular compartment and suspended plant material, as yet unfrozen, is excluded from the ice crystal lattices and becomes entrapped in the channels between ice crystals. These channels are filled with a limited volume of residual solution containing solutes that composed the original extracellular solutions. These have been excluded from the ice crystal lattice, and as they accumulate in the residual solution its freezing point is depressed. As the system is cooled still further, more water freezes into ice, further reducing the volume of the residual solution and increasing its solute burden (Grout and Morris 1987). This generates an increasingly negative water potential in the residual solution that results in significant osmotic water loss from the plant materials, described as cryodehydration.

The effective molarity of the residual solution, when in equilibrium with ice at a specified temperature, is determined by the relationship:

#T °C / 1.86 (osmoles litre^{-1}),
where #T °C is the depression of freezing point.

The ice/solution system (see Fig. 2) will result in continued cryodehydration until a eutectic temperature is reached, at which point solute crystallizes from solution, e.g. a sodium chloride/water system will solidify below −21.8 °C. If solutions contain more than one solute, then multiple eutectic points can be observed, with some solutes coming out of solution before others.

The function of controlled, slow cooling in cryopreservation is to allow cryodehydration to progress without intracellular freezing, removing water from cells to a point where their contained solutions will not form ice crystals when taken to the final cryogen temperature. Instead, they will undergo a transition to a non-crystalline or amorphous phase, known as a glass, by a process known as vitrification. At the temperatures involved in cryogenic storage, the presence of vitrified solutions inside the cells is essentially non-damaging. For a particular biological system there is, therefore, an optimal slow rate of cooling which is essentially the optimal rate of dehydration to allow for subsequent vitrification when the sample is immersed directly in liquid nitrogen. This is done from an intermediate temperature typically

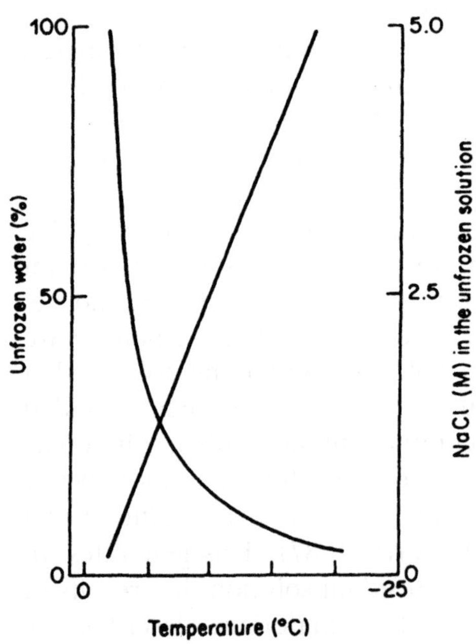

Fig. 2. Percentage of unfrozen water and the molarity of solute in unfrozen residual solutions of sodium chloride at different subzero temperatures. The initial molarity of the sodium chloride was 0.15 M. (Grout and Morris 1987)

between −35 and −40 °C. If the temperature cools below this latter level, and intracellular solutions are still relatively dilute, inevitably intracellular ice crystals will form (Meryman and Williams 1985), which are lethal (Grout 1991b) except under rare circumstances where they are very small and of a particular ice conformation (Rall et al. 1980; Boutron 1987). Obviously, intracellular freezing must be prevented if cryopreservation is to be successful. Ice intrusion from extracellular locations into the cytosol is very efficiently restricted when the plasma membrane is intact (Grout and Morris 1987) but will occur where any structural lesion exists (Asahina 1962; Mazur 1977; Steponkus 1984). Intracellular ice formation in intact cells also depends upon the extent of undercooling and cooling rate (Grout 1991b) and, typically, plant cells can undercool by 10 °C, whilst having a cytoplasmic equilibrium freezing point between −2 and −50 °C. In the presence of extracellular ice and at more rapid rates of cooling, there will be restricted time for osmotic water loss by cryodehydration. Consequently, the cytoplasm will undercool significantly, which increases the probability of intracellular nucleation forming lethal cytoplasmic ice.

Ice crystals will also have potentially lethal, disruptive effects when in an extracellular situation, whether outside tissues or between the cells of a tissue or organ. These effects will include:

- Mechanical stresses on material entrapped in the channels between crystals, causing deformations and tissue damage (Grout 1991b).
- Ice formed between cell walls and the shrinking protoplast that can adhere to the membrane surface generating adhesion energies sufficiently strong to result in lysis (Tandorf et al. 1987; Grout 1991b).
- Electrical interactions at membrane surfaces close to ice fronts (the Workman-Reynolds effect) resulting from the different solubilities of ions in the ice and liquid phases. This leads to perturbation of biological membranes (Steponkus 1984).
- Intracellular gas bubble formation that may occur under a number of circumstances and can be observed microscopically during cooling and warming of isolated plant protoplasts and unicellular algae (Grout 1991b). Chemical injuries related to lipid peroxidation and localised pH changes have been proposed as causes of "bubble-related" contact injury (Morris and McGrath 1981).

Further, potentially lethal effects will result from the hypertonicity of the residual solution. These include:

- Irreversible surface area reduction of the limiting cell membrane following osmotic shrinkage (Steponkus 1984; Grout and Morris 1987).
- An increase in the concentration of cytoplasmic solutes.
- Osmotic shrinkage of organelles.

Cells shrink in response to the increasing osmolarity of the residual solution and the initial reduction in surface area of the limiting membrane is achieved by lateral compression of components, followed by deletion of material from the plane of the membrane by simple folding, vesiculation or deletion (Steponkus 1984; Grout 1991b). If deletion of membrane material occurs beyond a critical limit, and is irreversible, then the membrane will be unable to contain the original isotonic volume when the cells are thawed and rehydrate, and lysis will result. Injuries may also result from changes im membrane permeability and fluidity associated with deletion of certain lipid and protein molecules.

When the extracellular medium is frozen, cytoplasmic macromolecules and cytoskeletal components will be exposed to increased intracellular solute concentrations, following cryodehydration, and will show differing sensitivities. Denaturation and deactivation of enzymes, for example, can result, as well as displacement of protein from membranous structures (Franks 1985; Grout 1991b). The stresses upon organelles will parallel those affecting the entire cell. Potentially damaging physical changes also occur on rewarming, particularly if this is slow compared to cooling. As temperature rises, water molecules diffuse from smaller ice crystals to larger ones, which grow, and increase the probability of strain in the biological material due directly to the presence of ice. This is known as migratory recrystallisation, and is fastest when small crystals initiated at low temperatures are stored at higher temperatures, which justifies, in part, the ultra-low temperatures used in cryopreservation.

To protect against the desiccation injuries resulting from cryodehydration, a range of chemical cryoprotectants are used as additives to the extracellular medium, and many of these will permeate to intracellular sites (Meryman and Williams 1985). The compounds used in plant tissue and organ cryopreservation are, in the main, those that have been proved successful with animal cells and tissues, and the most widely used are dimethyl sulphoxide, glycerol, sucrose and methanol. They are used at relatively high concentrations, commonly up to 2.5 M, and so may be expected to generate stresses due to toxicity and osmotic effects in their own right. The more suitable cryoprotectants are characterised by a relatively low toxicity and are significantly permeant to cells and tissues. Their colligative role depends upon an effective reduction in the water content of the cell, which, in turn, reduces the effective ion/solute concentrations that can occur as a result of cryodehydration. This is illustrated by sodium chloride/glycerol mixtures in which a slowly frozen 0.154M NaCl solution is concentrated by a factor of approximately 15 at −10°C. The addition of glycerol at a 1:1 ratio (by weight) will reduce the NaCl concentration to ca. ×9, and with a 9:1 addition this reduces even further to ca. ×3.5 at the same temperature. At slow cooling rates, and assuming that the cryoprotectant is equilibrated across the limiting membranes of the cells concerned, this colligative action also reduces the amount of cell shrinkage that will occur at any given temperature following extracellular freezing, thus reducing the types of mechanical injuries associated that have been discussed above.

Larger, non-permeating polymeric cryoprotectant compounds have also proved to be effective protectants for a limited range of material. Their mode of action may be, in part, colligative, as they become increasingly more concentrated in residual solutions as freezing progresses, and do have an effect on freezing point depression and the reduction of ion concentrations (Kartha 1985). They form supersaturated solutions of very high viscosity as the residual solution volume decreases, for they do not crystallise readily from aqueous solutions, and in this condition constrain a certain amount of water, which is rendered effectively unfreezable and can beneficially influence ice formation in the system. Polyethylene glycol (PEG), polyvinyl pyrrolidone (PVP) and hydroxyethyl starch (HES) are in this category, and it has been calculated that HES may prevent freezing of up to 50% of its own weight of water (Korber et al. 1982). A number of cryoprotectant compounds have other specific properties that may be of value, or concern, during cryopreservation. Dimethyl sulphoxide is also reported to cause a diversity of genetic and epigenetic effects on cells when used at high concentrations and with extended exposure times, and it has been used in transformation studies with leukaemic cell lines. However, there is no convincing evidence for genetic alterations due to cryoprotectant effects in cryopreservation experiments, where, typically, concentrations of protectant are relatively low, exposure times are short and reduced temperatures are likely to have an ameliorating effect. It has also been suggested that freezing damage is related, in part, to free radical effects, and that both dimethyl sulphoxide and glycerol provide an element of protection against these agents by acting as free radical scavengers (Benson 1990).

There is a clear benefit to be gained when attempting cryopreservation from exploiting any natural ability of the donor plants to cold-harden (Meryman and Williams 1985; Reed 1989). This may be done by taking explants from donor plants or in vitro cultures that have cold-hardened naturally in the field or been acclimated in the laboratory. In plants acclimated to low temperature, the lipid composition of the plasma membrane is modified in a way that reduces the transition of membrane lipids from lamellar to hexagonal forms during cryodehydration (Sugawara and Steponkus 1990; Webb and Steponkus 1990). In turn, this removes the major cause of osmotic non-responsiveness and destabilisation of the plasma membrane that is a primary injury in frozen/thawed material. During effective cold-hardening there is also significant accumulation of intracellular solutes, which commonly have cryoprotectant activity, such as glycine betaine, proline and soluble sugars (Koster et al. 1989). Attempts at acclimation of parent tissues should be made, therefore, with due regard to the physiology of the parent plant. If the donor tissue is an in vitro liquid culture, such as a suspension of somatic embryos, there may also be a benefit in attempting acclimation by osmotic stress, which has been successful in the cryopreservation of cell lines in suspension (Pritchard et al. 1986).

As experience with cryopreserved plant organ systems has accumulated, the concepts of sublethal injury and a phase of repair upon thawing have been given greater consideration (Benson and Withers 1987; Benson and Noronha-Dutra 1988; Benson 1990; Grout 1991b). It is probably a very exceptional circumstance when material is recovered from cryopreservation without cellular damage, however transient, and acclimation, pregrowth or other pretreatments, modified recovery media and culture conditions should all be considered in a practical situation (Benson and Hamill 1991; Grout 1991b).

To be confident that material recovered from cryogenic storage is not only viable but stable and unaltered in all important respects there are a wide range of assays that might be performed, depending upon the objectives of the storage system (Harding 1991; Potter and Jones 1991; Ward et al. 1993). These include:

- morphology
- protein electrophoresis
- secondary product synthesis
- karyotype analysis
- DNA restriction fragment analysis
- flow cytometry
- growth rates and yield
- enzyme activities
- gaseous evolution
- chromosome staining
- ribosomal RNA gene analysis

It is important that the material is characterised appropriately before freezing as well as on recovery, and that a realistic storage period is allocated for any given sample, after which it will be recovered and regrown. Subsequently, a second cycle of evaluation and cryopreservation can be undertaken. This is particularly important when the aim of the storage procedure is germplasm conservation, and the value of precise and comprehensive documentation at all stages of the procedure cannot be underestimated.

1.2.2 Vitrification by Direct Immersion in Cryogen

Vitrification is the phase transition of water from a liquid directly into a non-crystalline or amorphous phase, a glass, by an extreme elevation in viscosity during cooling (Fahy et al. 1984). Vitrification of any significant volume of pure water is rarely achieved and would require an ultra-rapid cooling rate, whereas solutions, such as those found within and surrounding cells during cryopreservation, vitrify at lower cooling rates as the solute concentration increases (Fahy et al. 1984; Angell and Sanapti 1987). With regard to cellular preservation, the conversion of cytoplasmic and extracellular solutions to a glass by vitrification following cooling avoids the damaging stresses associated with crystalline ice formation and cryodehydration (outlined above), and so increases viability following recovery from the cryogen. The preparative procedures for vitrification may themselves reduce viability, but the simple cooling process will produce little further viability loss. In the simplified phase diagram representing a typical solution that might be found in a biological context (Fig. 3), vitrification occurs as the curve representing the homogeneous nucleation temperature intersects that representing the glass

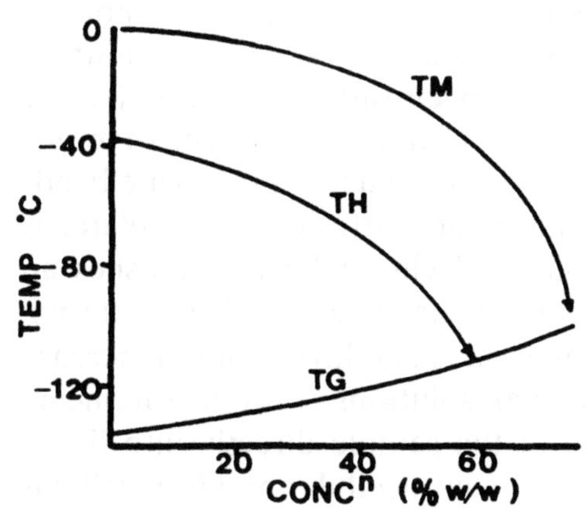

Fig. 3. A simplified phase diagram for a hypothetical compound with cryoprotectant activity, showing the conditions that can lead to vitrification. This would occur at the point where TH intersects TG. *TM* Melting point curve; *TH* homogeneous nucleation curve; *TG* glass transition temperature. (After Mackenzie and Luyet 1962)

transition temperature. Successful recovery of vitrified material is also dependent upon rapid thawing, as this ensures that the glass returns to an aqueous solution without recrystallisation, i.e. a transition from the vitreous to the crystalline phase, that might compromise viability (Fahy 1987; McFarlane and Forsyth 1987; Grout 1991a). A simple demonstration of recrystallisation can be made using transparent plastic straws (typically used for semen freezing) as sample vessels. A solution in the straw is transparent when successfully vitrified and will become visibly opaque due to ice crystal formation as it is slowly warmed (which would be inappropriate for successful cryopreservation). A significant practical advantage of vitrification for cryopreservation is that, once manipulated to be sufficiently concentrated to make the phase transition, vitrification of cellular solutions will occur at a cooling rate that can be achieved by direct immersion of a conventional sample vessels (straws, polypropylene ampoules) into liquid nitrogen. This removes the need for specialist equipment to control rate of cooling, and makes the technique more readily accessible to laboratories with resource limitations.

As in preservation by slow cooling, the key to successful vitrification is the effective increase of solutes in the cellular solutions, achieved by loading and/or removal of water. These changes have to be made within precise limits and the achievable levels of both mass and heat transfer will limit the size of the vessels, and the in vitro explants themselves, that can be successfully vitrified using current methodologies. Excised meristems, smaller shoot tips and somatic embryos are suited to the technique, as well as protoplasts and cells in suspension. The concentration of cellular solutes may be manipulated in one of two ways. The tissues may be incubated in hypertonic solutions, typically at low, non-freezing temperatures containing compounds of known cryoprotectant activity to protect against water stress injuries. Solutes will enter the cells in response to concentration gradients (solute loading) and water will be taken out, moving along a gradient of water potential. Alternatively, the explants may be encapsulated in alginate beads prior to vitrification, which is likely to moderate diffusion rates. Water is removed from the system by controlled air drying, which establishes an extreme gradient of water potential between the tissues and the external environment (see Table 1). Whichever method is employed should increase the concentration of solutes in intracellular solutions, without sacrificing viability, to the point where immersion of the sample directly into liquid nitrogen will bring about vitrification of both extracellular and intracellular solutions.

Prior to treatments outlined above, there is a clear benefit of treatment of the material to exploit any natural tendency towards cold-hardening (Reed

Table 1. Examples of in vitro plant cultures cryopreserved by vitrification

Dehydrated in solution		
Brassica campestris	Cell suspensions	Langis et al. 1989
Asparagus officinalis	Isolated cells Somatic embryos	Uragami et al. 1989
Secale cereale	Protoplasts	Langis and Steponkus 1990
Mentha sp.	Shoot tips	Towill 1990
Citrus sinensis	Isolated cells	Sakai et al. 1991
Trifolium repens	Apical meristems	Yamada et al. 1991a
Ribes sp.	Apical meristems	Reed 1992
Asparagus officinalis	Multiple bud clusters	Sakai et al. 1992
Chrysanthemum morifolium	Shoot tips	Schabel-Preikstas et al. 1992a
Solanum tuberosum	Shoot tips	Schabel-Preikstas et al. 1992b
Air dried with, or without[a], alginate encapsulation		
Solanum tuberosum	Shoot tips	Fabre and Dereuddre 1990
Asparagus officinalis	In vitro axillary buds[a]	Uragami et al. 1990
Daucus carota	Somatic embryos	Dereuddre et al. 1991
Melon	Somatic embryos[a]	Shimonishi et al. 1991
Euclyptus gunnii	Shoot tips	Monod et al. 1992
Pyrus sp.	Shoot tips	Scottez et al. 1992
Brassica napus	Microspore embryos (also)[a]	Uragami et al. 1993
Saccharum sp.	In vitro apices	Paulet et al. 1993

1989; Dereuddre et al. 1991). This involves appropriate cold acclimation of donor plants for the explant whether they be from the field, or growing in vitro, e.g. chrysanthemum meristems are excised from cuttings stored in the dark for 2 weeks at 5 °C (Schnabel-Preikstas et al. 1992a). A pregrowth period for the explant is also beneficial in most instances, whether the material is to be treated in solution or air-dried, e.g. excised meristems of *Trifolium repens* are cultured in medium supplemented with 1.2 M sorbitol for 2 days at 4 °C prior to applying the final vitrification solutions (Yamada et al. 1991a). In a protocol using alginate-coated subjects, encapsulated carrot somatic embryos were precultured in 0.3 M sucrose for 18 h. Abscisic acid has also been successfully used as a pregrowth treatment (Shimonishi et al. 1991). The effects of this type of pregrowth have not been investigated in detail, but it is probable that they include solute accumulation, reduction of isotonic water content and structural changes such as a reduction in individual vacuolar volumes (Pritchard et al. 1986; Reed 1989, 1992) and changes in basic membrane structure (Koster et al. 1989; Sugawara and Steponkus 1990; Webb and Steponkus 1990).

Vitrification solutions, and those used in pretreatments, commonly contain compounds which exhibit cryoprotectant activity when employed in slow freezing protocols, suggesting they have some protectant action against the stresses of cytoplasmic dehydration caused by the vitrification solution, or during air drying. Once thawed, the osmotic environment of the explants is gradually reduced to the original isotonic level, and concentration gradients adjusted to allow the protectant solutes accumulated within the cells to diffuse out and be dispersed. In successful cases of preservation, the intracellular solutes will become redistributed to the physiological state and the normal water status of the cells and organelles will be restored.

In protocols employing air drying, there is no further treatment with the cryoprotectant compounds following pregrowth treatments, and protection from desiccation injuries must be provided by a combination of:

- The effects of cold acclimation on donor material.
- The effects of the pregrowth period, which may include intracellular accumulation of solutes, e.g. sucrose, used in the preculture medium as well as responses to osmotic stresses.
- Natural tolerance to desiccation that is a property of many meristematic plant tissues.

Resistance to the stresses of cryopreservation can be acquired as a consequence of natural hardening during the winter season. In mulberry, Scots pine and pear, for example, excised vegetative buds collected from the field in the winter season (at maximum natural hardiness) can be progressively cooled into liquid nitrogen with no further pretreatment, preserving high levels of viability (Yakuma and Oka 1988; Kuosoka and Hohtola 1991; Oka et al. 1991). There are also instances of explants being successfully cryopreserved following direct immersion in liquid nitrogen, but with cryoprotectant treatments characteristic of a slow cooling protocol, rather than exposure to a vitrification solution. Examples include excised meristems of potato (Grout and Henshaw 1978), cassava, (Kartha et al. 1982) and rapeseed (Withers 1982). These explants were held in small volumes of the cryoprotectant solution, providing for a rapid cooling rate, and were frozen on the tip of a hyperdermic needle (Grout and Henshaw 1978) or as small droplets on an aluminium foil sheet (Kartha et al. 1982). There is also a reported instance of low survival of entire in vitro plantlets of sugarcane (2–3 cm in height) following air drying without preculture and direct immersion in liquid nitrogen (Bajaj 1990). Vitrification mechanisms are likely to be behind the success of such procedures.

References

Ammirato PV (1983) Embryogenesis. In: Evans DA, Sharp WR, Ammirato PV, Yamada Y (eds) Handbook of plant cell culture, vol 1: Techniques for propagation and breeding. Macmillan, New York, pp 82–123

Angell CA, Sanapti H (1987) Crystallisation and vitrification in cryoprotected aqueous systems. In: Pegg DE, Karow AM (eds) The biophysics of organ cryopreservation. Plenum Press, New York, pp 147–172

Asahina E (1962) Frost injury in living cells. Nature 196:445–446

Ashwood-Smith MJ, Farrant J (1980) Low temperature preservation in medicine and biology. Pitman Medical, Tunbridge Wells

Ashwood-Smith MJ, Grant E (1977) Genetic stability in cellular systems stored in the frozen state. In: Elliot K, Whelan J (eds) The freezing of mammalian embryos. Elsevier, Amsterdam, pp 251–268

Bajaj YPS (1990) Cryopreservation of germplasm of vegetatively propagated crops. Bull Soc Bot Fr Actual Bot 137:99–114

Bateson JM, Grout BWW, Lane S (1987) In: Ducote G, Jacob M, Simeon A (eds) Plant micropropagation in the horticultural industries. Belgium PTC Group, Arlon, pp 275–277

Belcher AR, Abbott AJ, Hall KC, Jackson MB (1987) Gaseous constituents of cultures in vitro. Report of the Long Ashton Res Stn 1986, AFRC, UK

Benson EE (1990) Free radical damage in stored plant germplasm. IBPGR, Rome

Benson EE, Hamill JD (1991) Cryopreservation and post-freeze molecular and biosynthetic stability in transformed roots of *Beta vulgaris* and *Nicotiana rustica*. Plant Cell Tissue Organ Cult 24:163–172

Benson EE, Noronha-Dutra AA (1988) Chemiluminescence in cryopreserved plant tissue cultures: the possible involvement of singlet oxygen in cryoinjury. Cryo Lett 9:120–131

Benson EE, Withers LAW (1987) Gas chromatographic analysis of volatile hydrocarbon production by cryopreserved plant tissue cultures: a non-destructive method for assessing stability. Cryo Lett 8:35–46

Bhojwani SS (ed) (1990) Plant tissue culture: applications and limitations. Elsevier, Amsterdam

Boutron P (1987) Non-equilibrium formation of ice in aqueous solutions: efficiency of polyalcohol solutions for vitrification. In: Pegg DE, Karow AM (eds) The biophysics of organ cryopreservation. Plenum Press, New York, pp 210–236

DeProff MP, Maenen L, Debergh P (1985) Carbon dioxide and ethylene evolution in the culture atmosphere of *Magnolia* cultured in vitro. Physiol Plant 65:375–379

Dereuddre J, Blandin S, Hassen N (1991) Resistance of alginate coated somatic embryos of carrot (*Daucus carota* L.) to desiccation and freezing in liquid nitrogen. 1. Effects of preculture. Cryo Lett 12:125–134

Dodds J (1991) In vitro methods for conservation of plant genetic resources. Chapman & Hall, London

Fabre J, Dereuddre J (1990) Encapsulation-dehydration: a new approach to cryopreservation of *Solanum* shoot tips. Cryo Lett 11:413–426

Fahy GM (1987) Biological effects of vitrification and devitrification. In: Pegg DE, Karow AM (eds) The biophysics of organ cryopreservation. Plenum Press, New York, pp 265–300

Fahy GM, McFarlane DR, Angell CA, Meryman HT (1984) Vitrification as an approach to cryopreservation. Cryobiology 21:407–426

Fay MF (1994) In what situations is in vitro culture appropriate to plant conservation? Biodiversity and conservation 3:176–183

Franks F (1985) Biophysics and biochemistry at low temperatures. Cambridge University Press, Cambridge

Gould JH, Murashige T (1985) Morphogenic substances released by plant tissue cultures. Plant Cell Tissue Organ Cult 4:298–342

Grout BWW (1990a) In vitro conservation of germplasm. In: Bhojwani SS (ed) Plant tissue culture: applications and limitations. Elsevier, Amsterdam, pp 394–411

Grout BWW (1990b) Genetic preservation in vitro. In: Nijkamp HJJ, Van der Plas LHW, Van Aartrijk J (eds) Plant cellular and molecular biology. Kluwer Dordrecht, pp 13–22

Grout BWW (1991a) The effects of ice during the cryopreservation of clinical systems. In: Fuller BJ, Grout BWW (eds) Clinical applications of cryobiology. CRC Press, Boca Raton, pp 81–94

Grout BWW (1991b) Cryopreservation of plant cells and organs. In: Zakri AH, Normah MT, Senawi MT, Abdul Karim AG (eds) Conservation of plant genetic resources through in vitro methods. FRIM/MNCPGR, Kuala Lumpur, pp 43–55

Grout BWW, Henshaw GG (1978) Freeze-preservation of potato shoot-tip cultures. Ann Bot 42:1227–1229

Grout BWW, Morris GJ (1987) Freezing and cellular organisation. In: Grout BWW, Morris GJ (eds) The effects of low temperatures on biological systems. Edward Arnold, London, pp 147–174

Harding K (1991) Molecular stability of the ribosomal RNA genes in *Solanum tuberosum* plants recovered from slow growth and cryopreservation. Euphytica 55:141–146

Hobbs PV (1974) Ice physics. Oxford University Press, Oxford

Karp A (1989) Can genetic instability be controlled in plant tissue cultures? Int Assoc Plant Tissue Cult Newsl 58:2–11

Karp A, Bright JP (1985) On the causes and origins of somaclonal variation. In: Miflin BJ (ed) Oxford surveys of plant molecular and cell biology, vol 2. Oxford University Press, Oxford, pp 199–234

Kartha KK (ed) (1985) Cryopreservation of plant cells and organs. CRC Press, Boca Raton

Kartha KK (1987) Advances in cryopreservation technology of plant cells and organs. In: Green CE, Somers DA, Hackett WP, Biesboer DD (eds) Plant tissue and cell culture. Alan R. Liss, New York, pp 447–458

Kartha KK, Leung NL, Mroginski LA (1982) In vitro growth responses and plant regeneration from cryopreserved meristems of cassava (*Manihot esculenta* Crantz). Z Pflanzenphysiol 107:133–140

Korber C, Scheiwe MW, Boutron P, Rau G (1982) The influence of hydroxyethyl starch on ice formation in aqueous solutions. Cryobiology 19:478–492

Koster KL, Steponkus PL, Lynch V (1989) Solute accumulation during cold acclimation of rye. Plant Physiol 89:S-26

Kuosoka T, Hohtola A (1991) Freeze preservation of buds from Scots pine trees. Plant Cell Tissue Organ Cult 27:89–93

Langis R, Steponkus PL (1990) Cryopreservation of rye protoplasts by vitrification. Plant Physiol 92:666–671

Langis R, Schnabel B, Earle ED, Steponkus PL (1989) Cryopreservation of *Brassica campestris* L. cell suspensions by vitrification. Cryo Lett 10:421–428

Larkin PJ (1987) Somaclonal variation, history, method and meaning. Iowa State J Res 61:393–434

Lyons JM, Raison JK, Steponkus PL (1979) The plant membrane in response to low temperature: an overview. In: Lyons JM, Graham D, Raison JK (eds) Low temperature stress in crop plants – the role of the membrane. Academic Press, London, pp 1–24

Mackenzie AP, Luyet BJ (1962) Electron microscope study of the structure of very rapidly frozen gelatin solutions. Biodynamica 9:47–69

Mazur P (1977) The role of intracellular freezing in the death of cells cooled at supra-optimal rates. Cryobiology 14:251–272

McFarlane DR, Forsyth M (1987) Devitrification and recrystallisation of glass-forming aqueous solutions. In: Pegg DE, Karow AM (eds) The biophysics of organ cryopreservation. Plenum Press, New York, pp 237–264

Meryman HT, Williams RJ (1985) Rasic principles of freezing injury to plant cells: natural tolerance and approaches to cryopreservation. In: Kartha KK (ed) Cryopreservation of plant cells and organs. CRC Press, Boca Raton, pp 13–48

Monod V, Poissonier M, Paques M, Dereuddre J (1992) Cryopreservation of shoot tips of in vitro plantlets of *Eucalyptus* after encapsulation and air dehydration. Cryobiology 29:737–738

Morris GJ, Grout BWW (1990) Technological synthesis. Philos Trans R Soc Lond B 326:687–692

Morris GJ, McGrath JJ (1981) Intracellular ice nucleation and gas bubble formation in *Spirogyra*. Cryo Lett 2:341–352

Nijkamp HJJ, Van der Plas LHW, Van Aartrijk J (eds) (1990) Progress in plant cellular and molecular biology. Kluwer, Dordrecht

Oka S, Yakuwa H, Sato K, Niino T (1991) Survival and shoot formation in vitro of pear winter buds cryopreserved in liquid nitrogen. HortScience 26:65

Paulet F, Engelmann F, Glaszmann JC (1993) Cryopreservation of apices of in vitro plantlets of sugarcane (*Saccharum* sp. hybrids) using encapsualtion/dehydration. Plant Cell Rep 12:525–529

Philips RL, Kaeppler SM, Peschke VM (1990) Do we understand somaclonal variation. In: Nijkamp HJJ, Van der Plas LHW, Van Aartrijk J (eds) Progress in plant cellular and molecular biology. Kluwer Dordrecht, pp 131–141

Pollard JW, Walker JM (eds) (1990) Methods in molecular biology, vol 6. Plant cell and tissue culture. Humana Press, New Jersey

Potter RH, Jones MGK (1991) Molecular analysis of genetic stability. In: Dodds JH (ed) In vitro methods for conservation of plant genetic resources. Chapman & Hall, London, pp 71–91

Priel W, Beck A (1991) Somatic embryogenesis in bioreactor culture. Acta Hortic 289:179–192

Pritchard HW, Grout BWW, Short KC (1986) Osmotic stress as a pregrowth procedure for cryopreservation 1. growth and ultrastructure of sycamore and soybean cell suspensions. Ann Bot 57:41–48

Rall WF, Reid DS, Farrant J (1980) Innocuous biological freezing during warming. Nature 286:511–514

Reed BM (1989) The effect of cold hardening and cooling rate on the survival of apical meristems of *Vaccinium* species frozen in liquid nitrogen. Cryo Lett 10:315–322

Reed BM (1991) Application of gas-permeable bags for in vitro cold storage of strawberry germplasm. Plant Cell Rep 10:431–434

Reed BM (1992) Cryopreservation of *Ribes* apical meristems. Cryobiology 29:740

Sakai A, Kobayashi S, Oiyama I (1991) Survival by vitrification of nucellar cells of navel orange (*Citrus sinensis* var. *brasiliensis* Tanaka) cooled to −196 °C. J Plant Physiol 137:465–470

Sakai A, Niino T, Yamada T, Kohmura H (1992) Cryopreservation by vitrification of mulberry meristems, meristematic callus of white clover and multiple bud cultures of asparagus. Cryobiology 29:747–748

Schnabel-Preikstas B, Earle ED, Steponkus PL (1992a) Cryopreservation of *Chrysanthemum* shoot tips by vitrification. Cryobiology 29:739

Schnabel-Preikstas B, Earle ED, Steponkus PL (1992b) Cryopreservation of potato shoot tips by vitrification. Cryobiology 29:747

Scottez C, Chevreau E, Godard N, Arnaud Y, Duron M, Dereuddre J (1992) Cryopreservation of cold-acclimated shoot tips of pear in vitro after encapsulation and dehydration. Cryobiology 29:691–700

Scowcroft WR (1984) Genetic variability in tissue culture: impact on germplasm conservation and utilisation. IBPGR, Rome

Shimonishi K, Ishikawa M, Suzuki S, Oosawa K (1991) Cryopreservation of melon somatic embryos by a dessication method. Jpn J Breed 41:347–351

Steponkus PL (1984) Role of the plasma membrane in freezing injury and cold acclimation. Annu Rev Plant Physiol 35:543–584

Sugawara Y, Steponkus PL (1990) Effect of cold acclimation and modification of the plasma membrane lipid composition on lamellar to hexagonal phase transitions in rye protoplasts. Cryobiology 27:667

Tandorf S, McGrath JJ, Olien CR (1987) On the adhesive interaction between ice and cell-size liposomes. Cryo Lett 8:322–328

Taylor MJ (1987) Physico-chemical principles in low temperature biology. In: Grout BWW, Morris GJ (eds) The effects of low temperatures on biological systems. Edward Arnold, London, pp 3–71

Terzi M, Loschiavo F (1990) Somatic embryogenesis. In: Bhojwani SS (ed) Plant tissue culture: applications and limitations. Elsevier, Amsterdam, pp 54–66

Towill LE (1990) Cryopreservation of isolated mint shoot tips by vitrification. Plant Cell Rep 9:178–180

Uragami A, Sakai A, Nagai M, Takahashi TA (1989) Survival of cultured cells and somatic embryos of *Asparagus officinalis* L. cryopreserved by vitrification. Plant Cell Rep 8:418–421

Uragami A, Sakai A, Nagai M (1990) Cryopreservation of dried axillary buds from plantlets of *Asparagus officinalis* grown in vitro. Plant Cell Rep 9:328–331

Uragami A, Lucas MO, Ralambosoa AJ, Renard M, Dereuddre J (1993) Cryopreservation of microspore embryos of rapeseed (*Brassica napus* L.) by dehydration in air with or without alginate encapsulation. Cryo Lett 14:83–91

Ward ACW, Benson EE, Blackhall NW, Cooper-Bland S, Powell W, Power JB, Davey MR (1993) Flow-cytometric assessments of ploidy stability in cryopreserved dihaploid *Solanum tuberosum* and wild *Solanum* species. Cryo Lett 14:145–153

Webb MS, Steponkus PL (1990) Dehydration-induced hexagonal phase formation in phospholipid bilayers. Cryobiology 27:666–667

Wilson J (1987) Chilling injury in plants. In: Grout BWW, Morris GJ (eds) The effects of low temperatures on biological systems. Edward Arnold, London, pp 271–292

Withers LA (1982) The development of cryopreservation techniques for plant cell, tissue and organ cultures. In: Fuijwara A (ed) Plant tissue culture 1982. Japanese Association for Plant Tissue Culture, Tokyo, pp 793–794

Withers LA (1985) Cryopreservation and storage of germplasm. In: Dixon RA (ed) Plant cell culture – a practical approach. IRL Press, Oxford, pp 169–192

Withers LA (1987) The low temperature preservation of plant cell, tissue and organ cultures and seed for genetic conservation and improved agricultural practice. In: Grout BWW, Morris GJ (eds) The effects of low temperatures on biological systems. Edward Arnold, London, pp 389–409

Yakuma H, Oka S (1988) Plant regeneration through meristem culture from vegetative buds of mulberry (*Morus bombycis* Koidz.) stored in liquid nitrogen. Ann Bot 62:79–82

Yamada T, Sakai A, Matsumura T, Higuchi S (1991a) Cryopreservation of apical meristems of white clover (*Trifolium repens* L.) by vitrification. Plant Sci 78:81–87

Yamada T, Sakai A, Matsumura T, Higuchi S (1991b) Cryopreservation of apical meristems of white clover (*Trifolium repens* L.). Plant Sci 73:111–116

2 Minimal Growth Storage

B.W.W. Grout

Minimal growth storage is the most direct way of restricting growth and development of in vitro materials and is most usually applied to differentiated plantlets and developing meristem cultures. It may also be applied to larger callus masses. At its simplest, lowering the environmental temperature of cultures below their active growing temperature is used to significantly extend the interval between subcultures. Care must be taken to avoid temperatures below freezing or levels where other temperature-related injuries might occur, e.g. chilling injury (Lyons et al. 1979; Wilson 1987). Despite the genetic stability inherent in organised plant structures, such as meristems, there are risks to DNA structure, integrated metabolism and viability under conditions of slow growth such as resulting from the oxidative activity of free radicals (Benson 1990). Non-lethal effects can also result from this type of damage and should be screened for by systematic evaluation of aspects of structure and function before, and after, storage (Potter and Jones 1991; Withers 1991). Minimal growth storage provides a relatively low level of stability and protection for in vitro subjects, as increase in biomass, development and metabolism are slowed rather than halted. However, the technique has the benefits of being inexpensive and requiring little, if anything, in the way of specialised equipment or expertise. It is readily accessible to workers concerned with experimental plant biology, plant improvement, or conservation per se, that have the basic facilities for in vitro culture and also the reduction of environmental temperature, often by as little as 10 °C. Suitable storage conditions often require little more than a standard refrigerator or cooled laboratory incubator, with minimal, or in some cases, no lighting. Where large numbers of individual cultures are to be stored, purpose-built cool rooms with ancillary lighting may be appropriate. Achievable, and desirable, storage periods before regeneration and re-evaluation of the stored material must necessarily be relatively short, with 12 months between subculture intervals being a realistic goal for many species (see Table 1).

The storage temperature is chosen not only with regard to the optimal growth temperature but also to the susceptibility to chilling injuries of the species concerned. Typically, the most suitable storage temperature will be between 1 and 10 °C for temperate species that are maintained in vitro

Table 1. Examples of species which have been stored in vitro as shoot cultures or plantlets by reducing incubation temperature

Species	Storage temp (°C)	Subculture interval (wks)	Reference
Storage for up to 1 year			
Trifolium repens	5	40	Bhojwani 1981
Xanthosoma brasiliense	6	8	Staritsky 1980
Vitis rupestris	4–12	24	Withers 1982
Vitis rupestris	9.5	26	Barlas and Skeene 1983
Prunus sp.	−3	40	Marino et al. 1985
Allium porrum	7	24	Dore 1988
Vitis rupestris	9–12	44	Galzy and Compan 1988
Euclyptus citriodera	10	36	Mascarhenas and Agrawal 1991
Coffea arabica	20	26	Bertrand-Desbrunais et al. 1992
Malus domestica	4	36	Orlikowska 1992
Storage for 1–2 years			
Malus domestica	4	52	Lundergen and Janick 1979
Ipomea batatas	22	52	Alan 1979
Trifolium repens	2–6	60	Cheyne and Dale 1980
Lolium multiflorum	2–4	52	Dale 1980
Solanum sp.	6–12	52	Westcott 1981a
Manihot esculentum	20	72	Withers 1982
Musa sp.	5	56	Banerjee and DeLanghe 1985
Dioscorea rotundata	18/22	52	Ng and Hahn 1985
Pyrus communis	4	72	Wanas et al. 1986
Allium sativum	4	64	El-Gizawy and Ford-Lloyd 1987
Saussurea lappa	5	52	Arora and Bhojwani 1989
Fragaria sp.	4	60	Reed 1991
Populus alba × *P. grandidentata*	4	104	Son et al. and 1991
Storage longer than 2 years			
Colocasia esculentum	9	156	Zandvoort and Staritsky 1986
Xanthomoas sp.	13	156	Zandvoort and Staritsky 1986
Colocasia esculenta	9	156	Bessembinder et al. 1993

between 20 and 27 °C. For species normally maintained in culture at higher temperatures, and those that are sensitive to chilling injuries (Staritsky 1986; Withers 1986; Mascarenhas and Agrawal 1991; Ng and Ng 1991), a more elevated storage temperature will be appropriate. The effectiveness of the system can, in some instances, be enhanced by various strategies including alternating day and night temperatures, e.g. 12/6 °C day/night for potato (Westcott 1981a), reducing the light levels, e.g. with *Musa* (Banerjee and DeLanghe 1985) or transferring the cultures to darkness, e.g. with *Populus*

(Chun and Hall 1986). Inhibitory levels of growth regulators may also be added to the culture medium, e.g. abscisic acid, dimethyl succinamic acid or CCC in *Solanum* (Westcott 1981a,b; Mix 1982; Withers 1986; Ng and Ng 1991).

Further, a level of controlled osmotic stress, using mannitol or sucrose, might be imposed on the plantlets, e.g., in *Solanum* (Westcott 1981b; Espinoza et al. 1984), in *Ipomea* (Ng and Hahn 1985), in *Actinidia* (Monette 1986) and in *Cinchona* (Hunter 1986).

Those familiar with the behaviour of plant material cultured in vitro will be aware of the significant variations that can occur as a result of species or varietal differences and, inexplicably, between laboratories. Consequently, the information contained in this chapter provides guidelines towards developing an optimal storage protocol in many instances, rather than a prescriptive methodology. The objectives of attempts at the successful minimal growth storage of in vitro-cultured plant materials must be to:

Objectives

- Devise an effective and reproducible protocol for minimising the growth and development of the plant materials.
- Avoid induction of abnormal patterns of growth and development during storage that might compromise subsequent normal growth.
- Preserve the viability of the stored population of in vitro cultures at a level that will avoid genetic selection and maintain the genetic diversity of the material in question.
- Ensure that the protocol allows for periodic growth under standard conditions in between periods of prolonged storage.

2.1 Growth at Reduced Temperature

Equipment and material

- Standard culture medium in suitable containers, with non-perishable closures. To ensure successful, extended culture, the vessels must be robust, and any closures and seals that are necessary should not deteriorate during the planned storage duration. There are handling advantages to the use of semi-solid medium, but liquid may be employed, although it is susceptible to evaporation under certain conditions.
- A secure culture environment with suitable, controlled temperature and with reduced lighting, as appropriate. Typically, the light intensity can be reduced by 60% from the standard growth level, e.g. 1000 lx and a 16-h photoperiod).

Procedures

1. Prepare culture vessels with the appropriate standard medium. An increased volume of medium may be used to prolong the storage period.
2. Establish fresh explants from active, in vitro material growing at the optimal temperature. Explants must be selected systematically from the entire population of cultures available, to ensure that there is no selection that might bias the genetic make-up of the stored population such that it is no longer representative of the original material.
3. Characterise the cultures in appropriate terms e.g., morphology, growth rates, isoenzyme profile to allow for subsequent evaluation of material recovered from storage.
4. Incubate the cultures to be stored at their optimal temperature until active growth is established.
5. Transfer the cultures to the selected storage temperature (see introductory paragraphy this Chap.). A range of 4–10 °C is likely to be the appropriate range for cultures grown at 20–27 °C, and 15–20 °C for those normally grown at temperatures higher than this. The optimal storage temperature for the circumstances involved will be determined by empirical investigation, and will depend upon the storage period required.
6. Establish a maintenance-level light regime, with a 16-h photoperiod at 500–1000 lx being widely applicable. Complete darkness may be applicable in some cases, but may lead to a problematic level of etiolation of plantlets. The most appropriate light level will be determined by experience.
7. Inspect cultures weekly, until satisfactory conditions have been confirmed, to check for contaminations and deterioration. Thereafter, 2-weekly inspections will suffice.
8. At a pre-selected point the cultures should be taken from the storage conditions, transferred to fresh growth medium and returned to optimal growth conditions. Subsequent performance is monitored and the cultures are compared with the original characterisation to confirm stability.

2.2 The Addition of Inhibitory Growth Regulators

Materials

- The choice of inhibitory growth regulators will necessarily depend upon circumstances, and abscisic acid, N-dimethylaminosuccinamic acid and CCC have all been used to effect.

Procedures

1. Incorporate the growth regulator into fresh growth medium at the required concentration, and use this medium to established new cultures (see previous procedure section). Optimal levels will be determined by experience, but suggested starting concentration ranges for these particular compounds are:

abscisic acid	5–50 mgl^{-1}
CCC (Cycocel)	2–10 mgl^{-1}
N-dimethylaminosuccinamic acid	5–50 mgl^{-1}

2. Maintain the new cultures under optimum conditions for 2–3 weeks to establish growth, although the inhibitory effect of the growth regulator will limit obvious signs of growth and development.

3. Otherwise as under Section 2.1.

2.3 The Addition of Osmotic Regulators

Materials

- The use of mannitol or sorbitol is recommended, as they are relatively metabolically inert, where the intention is to minimise growth by imposing a level of osmotic stress on the cultures.

Procedures

1. Incorporate the selected osmotic inhibitor into fresh growth medium. A recommended concentration, before attempts at optimisation, would be in the range 3–6% w/v. New cultures are established on this medium.

2. Maintain the new cultures under optimum conditions for 2–3 weeks to establish growth, although the inhibitory effect of the osmotic agent will limit obvious signs of growth and development.

3. Otherwise as under Section 2.1.

2.4 Combined Treatments

Whilst little systematic work has been published, there is evidence, supported by common sense, suggesting that combinations of the treatments outlined above may be of benefit, e.g. reduced temperature together with a growth inhibitor have been used with success for *Solanum tuberosum* (Westcott 1981b).

Investigation of the potential of combined treatments rests with the individual worker, and the safest general approach is to evaluate the effects of each of the individual restrictive treatments before attempting to use them in combination.

References

Alan JJ (1979) Tissue culture storage of sweet potato germplasm. PhD Thesis, University of Birmingham, UK

Arora R, Bhojwani SS (1989) In vitro propagation and low temperature storage of *Sassurea lappus* CB Clarke – an endangered medicinal plant. Plant Cell Rep 8:44–47

Banerjee N, DeLanghe E (1985) A tissue culture technique for rapid clonal propagation and storage under minimal growth conditions of *Musa* (banana and plantain). Plant Cell Rep 4:351–354

Barlass M, Skeene KGM (1983) Long-term storage of grape in vitro. Plant Genet Resour Newsl 53:19–21

Benson EE (1990) Free radical damage in stored plant germplasm. IBPGR, Rome

Bertrand-Desbrunais A, Noirot M, Charrier A (1992) Slow growth in vitro conservation of coffee (*Coffea* sp.) 2. Influence of reduced concentration of sucrose and low temperature. Plant Cell Tissue Organ Cult 31:105–110

Bessembinder JJE, Staritsky G, Zandvoort EA (1993) Long-term in vitro storage of *Colocasia esculentum* under minimal growth conditions. Plant Cell Tissue Organ Cult 32:121–127

Bhojwani, SS (1981) A tissue culture method for propagation and low temperature storage of *Trifolium repens* genotypes. Physiol Plant 52:187–190

Cheyne VA, Dale PJ (1980) Shoot tip culture in forage legumes. Plant Sci Lett 19:303–309

Chun YW, Hall RB (1986) Low temperature storage of in vitro cultured hybrid poplar, *Populus alba* x *P. grandidenta* plantlets. Abstr VI International Congress of Plant Tissue & Cell Culture. University of Minnesota, Minneapolis, p 13

Dale PJ (1980) A method for in vitro storage of *Lollium multiflorum* Lam. Ann Bot 45:497–502

Dodds J (1991) In vitro methods for conservation of plant genetic resources. Chapman & Hall, London

Dore C (1988) Multiplication végétative et conservation in vitro chez le poireau (*Allium porrum* L.). Agronomie (Paris) 8:509–511

El-Gizawy AM, Ford Lloyd BV (1987) An in vitro method for the conservation and storage of garlic (*Allium sativum*) germplasm. Plant Cell Tissue Organ Cult 9:147–150

Espinoza N, Estrada R, Tovar P, Bryan J, Dodds JH (1984) Tissue culture micropropagation, conservation and export of potato germplasm. Specialised Technology Document 1, CIP, Lima

Galzy R, Compan D (1988) Growth and nutrition of grapevine during in vitro long-term storage. Plant Cell Tissue Organ Cult 13:229–237

Hunter CS (1986) In vitro propagation and germplasm storage of *Cinchona*. In: Withers LA, Alderson PG (eds) Plant tissue culture and its agricultural applications. Butterworths, London, pp 291–301

Lundergen C, Janick J (1979) Low temperature storage of in vitro apple shoots. HortScience 14:514

Lyons JM, Raison JK, Steponkus PL (1979) The plant membrane in response to low temperature: an overview. In: Lyons JM, Graham D, Raison JK (eds) Low temperature stress in crop plants – the role of the membrane. Academic Press, London, pp 1–24

Marino G, Rosati P, Sagrati F (1985) Storage of in vitro cultures of *Prunus* rootstocks. Plant Cell Tissue Organ Cult 5:73–78
Mascarenhas AF, Agrawal DC (1991) Prospects for in vitro conservation of tree germplasm. In: Zakri AH, Normah MN, Senawi MT, Abdul Karim AG (eds) Conservation of plant genetic resources through in vitro methods. FRIM/MNCPGR, Kuala Lumpur, pp 231–252
Mix G (1982) In vitro preservation of potato materials. Plant Genet Resour Newsl 51:6–8
Monette PL (1986) Cold storage of kiwi fruit shoot tips in vitro. HortScience 11:1203–1205
Ng SYC, Hahn SK (1985) Application of tissue culture to tuber crops at IITA. In: Anon (ed) Proceedings of the Inter-Centre Seminar on International Agricultural Research Centers and Biotechnology. IRRI, Philippines, pp 27–40
Ng SYC, Ng NQ (1991) Reduced-growth storage of germplasm. In: Dodds JH (ed) In vitro methods for conservation of plant genetic resources. Chapman & Hall, London, pp 11–39
Orlikowska T (1992) The effects of in vitro storage at 4 °C on survival and proliferation of two apple rootstocks. Plant Cell Tissue Organ Cult 31:1–7
Potter RH, Jones MGK (1991) Molecular analysis of genetic stability. In: Dodds JH (ed) In vitro methods for conservation of plant genetic resources. Chapman & Hall, London, pp 71–91
Reed BM (1991) Application of gas-permeable bags for in vitro cold storage of strawberry germplasm. Plant Cell Rep 10:431–434
Son SA, Chun YW, Hall RB (1991) Cold storage of in vitro cultures of hybrid poplar shoots (*Populus alba* L. x *P. grandidentata* Michx). Plant Cell Tissue Organ Cult 27:161–168
Staritsky G (1980) In vitro storage of aroid germplasm. Plant Genet Resour Newsl 42:25–27
Staritsky G (1986) In vitro conservation of aroid germplasm at reduced temperatures and under osmotic stress. In: Withers LA, Alderson PG (eds) Plant tissue culture and its agricultural applications. Butterworths, London, pp 277–284
Wanas WH, Callow JA, Withers LA (1986) Growth limitation for the conservation of pear genotypes. In: Withers LA, Alderson PG (eds) Plant tissue culture and its agricultural applications. Butterworths, London, pp 285–290
Westcott RJ (1981a) Tissue culture storage of potato germplasm.1. Minimal growth storage. Potato Res 24:331–342
Westcott RJ (1981b) Tissue culture storage of potato germplasm.2. Use of growth retardants. Potato Res 24:343–352
Wilson J (1987) Chilling injury in plants. In: Grout BWW, Morris GJ (eds) The effects of low temperatures on biological systems. Edward Arnold, London, pp 271–292
Withers LA (1982) Storag of plant tissue cultures. In: Withers LA, Williams JT (eds) crop genetic resources – the conservation of difficult material. IBPGR, Rome, pp 49–82
Withers LA (1986) In vitro approaches to the conservation of plant genetic resources. In: Withers LA, Alderson PG (eds) Plant tissue cutlure and its agricultural applications. Butterworths, London, pp 261–276
Withers LA (1991) Tissue culture in the conservation of plant genetic resources. In: Zakri AH, Normah MT, Senawi MT, Abdul Karim AG (eds) Conservation of plant genetic resources through in vitro methods. FRIM/MNCPGR, Malaysia, pp 1–18
Zandvoort EA, Staritsky G (1986) In vitro gene banks of tropical aroids – research of storage conditions. Abstr VI International Congress of Plant Tissue & Cell Culture. University of Minnesota, Minneapolis, 426 p

3 Cryopreservation of Protoplast, Suspension and Callus Cultures

B.W.W. Grout

This chapter deals with cryopreservation of the simplest types of in vitro culture, at least in terms of differentiation and organisation, and also contains general information concerning aspects of cryopreservation which are relevant to other subjects in this volume, which will be cross-referenced elsewhere.

Isolated plant protoplasts are most commonly liberated from tissues, or cultured cells in vitro, by enzymic digestion of the cell walls (Power and Davey 1990). This occurs in a solution hypertonic to the parent tissues, to induce plasmolysis and rupture plasmodesmatal connections, so that spherical protoplasts, each representing the contents of a single cell, constitute the bulk of the released population. This hypertonic situation compensates for the absence of wall pressure and prevents protoplast lysis. Mechanical disruption of tissue has also been used to liberate protoplasts, although producing very low yields due to the inefficiency of the process and the high probability of physical damage (Cocking 1972). A population of isolated protoplasts will not be representative of the parent tissues, as senescent and highly vacuolate cells, particularly, are susceptible to the handling procedures and respond by rupturing. Viable protoplasts regenerate cell wall materials within a few hours of isolation, which destroys their most valuable property, notably access to an unencumbered plasmalemma surface. A technique for storage is, therefore, required that not only protects viability but also suspends regenerative development, particularly the production and accumulation of materials at the protoplast surface (Steponkus 1985; Bajaj 1988, 1989).

Suspension cultures comprise various proportions of single cells and cell aggregates of differing sizes in liquid medium. There is a considerable diversity of structure in most suspension cultures, with differences in size, the extent of vacuolation, and cell wall structure and thickness being the most apparent. There are also differences with regard to cell position in the mitotic cycle, as very few suspensions are synchronous in that regard. This degree of diversity is sufficient to ensure that any successful cryopreservation protocol will allow recovery of only a portion of the cell population (Withers 1990). The assumption, from a conservation point of view, is that the cell population is essentially clonal and that such limited genetic diversity as does exist (see Grout, Chap. 1, this Vol.) is uniformly represented throughout the phenotypes of the population.

Callus cultures have at least the range of structural and genetic diversity seen in suspension cultures, and are grown on medium solidified with agar or similar gelling agents. They grow as relatively large cell masses, many times larger than the aggregates typical of a suspension culture, and so pose significant problems of mass and heat transfer with regard to cryodehydration and cooling during preservation. Consequently, the partial survival of a callus cell population following cryopreservation is similar to that of a suspension culture in response to any particular protocol (Finkle et al. 1983).

Historically, cryopreservation using relatively slow cooling rates has been successful with the above categories of material, and the technique has undergone considerable refinement (Kartha 1985; Steponkus 1985; Withers 1985a,b; Grout and Morris 1987; Withers 1987; Withers 1990; Dodds 1991). Vitrification has recently also been shown to be successful for a limited range of suspension and protoplast cultures (see Towill, Chap. 7, this Vol.). There are a large number of commercially available machines that provide the required slow cooling rates over the 0 to −80 °C range for these biological samples. The use of such equipment provides reproducibility in the critical parts of the process. The generalised cooling procedure is essentially that discussed in Chapter 1, this volume, cooling at between 0.5 and 1.0 °C per min over the 0 to −40 °C range, prior to direct plunging of the sample into liquid nitrogen (practical details are given below). Where resources are limited, simple cooling apparatus that provides limited control over cooling rate can be assembled from standard laboratory items. A series of improvised structures are shown in Fig. 1, which can be valuable in simple investigations, but they rely heavily on the skill and ingenuity of the investigator to achieve anything like a linear cooling rate over the required temperature range. A more sophisticated apparatus is shown in Fig. 2, which will reliably reproduce the typical cooling rates of successful protocols (Withers and King 1980).

The protocols given in this and subsequent chapters offer a strong likelihood of success for the investigator but may, on occasion, require manipulation of one or more of the potential variables within the protocol. The areas where adjustments can be made (see Grout, Chap. 1, this Vol.) are:

- Duration and nature of pretreatments
- nature and concentration of protectants
- incubation time in protectant
- cooling rate

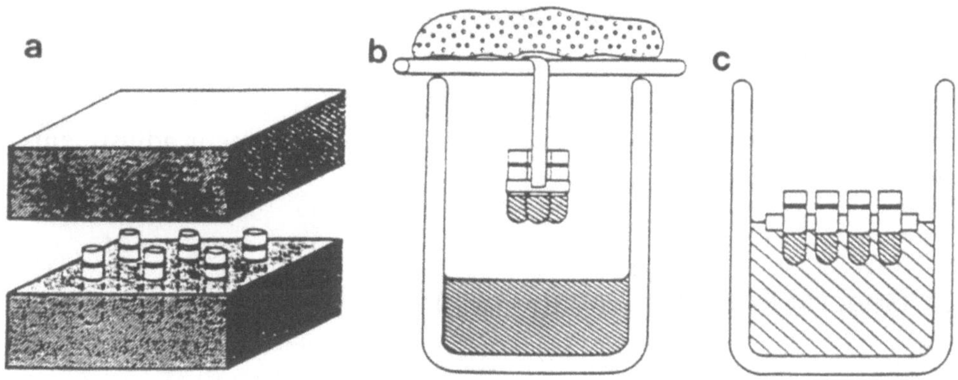

Fig. 1a–c. Simple apparatus to control cooling rate. **a** Vials are placed in an insulated container (conveniently made from expanded polystyrene) which, in turn, is placed inside a −80°C refrigerator. Cooling rate depends upon the nature and dimensions of the material making up the vial holder, and the temperature of the refrigerator environment. **b** Vials are suspended over liquid nitrogen, with as little convection movement as possible. Cooling rate is determined by the height above the liquid surface and can be adjusted as required by lowering or raising the vials. **c** In this construction vials are trapped in a polystyrene raft and floated on the surface of cooled alcohol, the cooling rate depending upon the alcohol temperature. (After Withers 1985b)

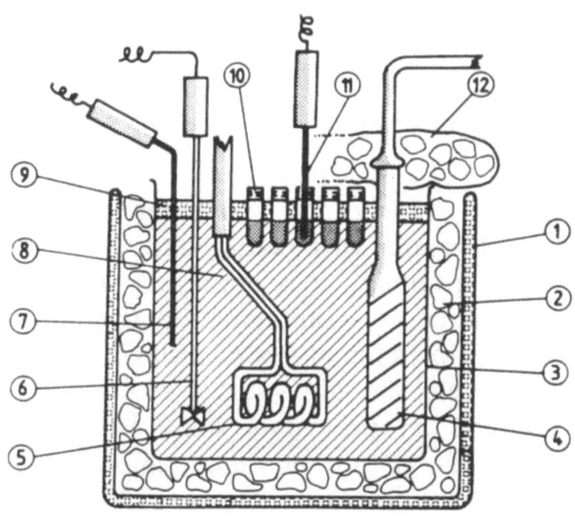

Fig. 2. A controlled-rate freezing apparatus designed by Withers and King (1980). The bath cools at a rate dependent upon the setting of the dip cooler and the volume of methanol. This apparatus can provide a holding period at a selected intermediate temperature. *1* Plastic container; *2* polystyrene insulation; *3* glass beaker; *4* dip-cooler; *5* heating coil; *6* stirrer; *7* temperature probe to regulate dip-cooler; *8* methanol; *9* raft of sheet polystyrene to hold vials and provide insulation; *10* vials; *11* thermocouple to monitor dummy sample temperature; *12* polythene chips in a bag for further insulation. (Withers and King 1980)

- intermediate temperature and holding time
- recovery conditions.

Optimisation of survival will almost certainly involve minor adjustment of several of these factors.

3.1 General Practical Considerations

If cryogenic storage, and recovery, is to be safe (for both workers and stored tissues) and successful, the following practical aspects of the laboratory procedures must be considered:

a) Maintenance of sterility and handling of reagents.

b) Sampling the original population before storage.

c) Characterisation and documentation.

d) Replication of samples in storage.

e) Hazards of working with liquid nitrogen.

f) Handling and performance of the storage system.

a) Maintenance of Sterility and Handling of Reagents. The material being stored will, almost always, be established as an in vitro culture under aseptic conditions. To maintain this sterile environment, it is important that all manipulations that might expose the tissues to contamination from the external atmosphere are carried out within the confines of a sterile work station. All freezing vials, pipettes, dishes etc. should be autoclaved or from freshly opened, pre-sterilised packs. The cryoprotectant solutions, obviously, must also be sterile but, given the nature of their components, autoclaving is generally not recommended. For small volumes, a filter unit attached to a hypodermic syringe will suffice, and for larger volumes, vacuum-assisted filter units with working volumes of 100ml upwards are recommended. The filters themselves should be resistant to compounds such as dimethyl sulphoxide (DMSO) at the concentrations used, and have a 2 μ pore size.

Where possible, cryoprotectant solutions should be freshly made, and can be stored in a refrigerator for 24h. The compound that poses particular difficulties is DMSO, which has a melting point of 18.4 °C and consequently, the undiluted stock solution will solidify on cool days in the laboratory. This should be avoided by appropriate storage conditions. DMSO should be stored in the dark, and in a glass container, as it attacks many plastics at higher concentrations. It has an unpleasant odour in concentrated solution and is reported to have mutagenic, teratogenic and carcinogenic properties. Consequently, it should be handled as a hazardous compound and all users should be trained in the appropriate handling and first-aid measures. If undiluted DMSO comes into contact with the skin, an unpleasant taste in the mouth can be expected within an hour or so unless the spill is rapidly washed away. This rapid permeation can facilitate uptake of other materials that normally are not permeant, and so present an additional potential hazard to the user. Glycerol, also widely used as a protectant, is often difficult to handle due to its high viscosity, and it is usually more convenient to measure it out by weight than by volume. Cryoprotectant mixtures should be adjusted to the pH of the standard growth medium *before* sterilisation.

b) Sampling the Original Population Before Storage. In order to maintain maximum genetic diversity and to avoid selection, care must be taken to ensure that the sample of explants or cells taken into storage is wholly representative of the original population. This should be done by applying a random sampling technique to the original population and, even in the simplest of cases, it may be prudent to consult an expert in the field to help construct the sampling strategy (see also next paragraph).

c) Characterisation and Documentation. The original population of donor plants/explants/cells should be characterised in appropriate ways, with regard to the likely end uses of the material. Characterisation may involve qualitative and/or quantitative assessment of aspects of morphology, physiology, biochemistry and genetics. These recorded data provide two points of reference for the investigator. Firstly, they allow comparison of the sample selected for storage with the parent population, to ensure that it is representative. Secondly, they allow the population recovered from storage to be evaluated for any alterations resulting from the storage process (see Harding and Benson, Chap 8, this Vol.).

Documentation must be the definitive record of the characteristics of the original population and the sample selected which, in turn, are the expected characteristics of the stored material. It is also the permanent record of complete and precise details of the storage and recovery procedures as they were carried out. These documents should be duplicated, and the copies held in separate, secure locations.

d) Replication and Placement in Storage. Once the explants/cells have been treated with cryoprotectant, it is important that they are transferred to the cryovials or foil packets in such a way that each vial/packet is representative of the treated population. Wherever possible, the vials/packets should be evenly distributed between several storage vessels and, if possible, between locations in the individual storage refrigerators.

e) Hazards Associated with Liquid Nitrogen. Liquid nitrogen is a markedly hazardous material with risks for both the direct user and those in the immediate vicinity. It is essential that users are fully trained in handling procedures by their local safety officer, including the use of suitable safety equipment and necessary first-aid procedures in case of accident.

Any area in which liquid nitrogen is to be used or stored must be well ventilated to reduce the risk of gaseous nitrogen build-up. Small volumes of the liquid convert into large volumes of gas (17 ml converts to more than 20 l) and the resulting local dilution of oxygen can cause drowsiness and, in extreme cases, asphyxiation. When large volumes of bulk liquid nitrogen are maintained in a relaitvely confined space, e.g. where there are many large storage vessels holding samples and bulk reserves of liquid nitrogen in a small laboratory, then an alarm monitor sensitive to the atmospheric level of gaseous nitrogen is advised.

Protective clothing, including a face shield, cryogloves and a suitable protective apron, should always be worn when working with liquid nitrogen. Vials and other containers of material stored in liquid nitrogen should always be handled with long forceps/tongs, and care should be taken when thawing cryovials, as occasionally they explode due to seepage of liquid nitrogen into the vial during storage.

f) Handling and Performance of the Storage System. A wide range of liquid nitrogen storage vessels for biological samples are commercially available, and those that store vials/packets/straws in the liquid phase of the nitrogen are to be recommended. The samples will be held in one of a range of internal systems that include clip-on canes for vials, canisters for straws and small boxes or racks in drawers. Whichever system is used, it is vital to ensure that transfers in and out of the system are made as rapidly as possible, and that no material in the storage vessel is allowed to warm significantly as a result of being lifted, temporarily, from the liquid cryogen.

To gain an appreciation of rewarming in this context it is a worthwhile exercise to fill a cryovial with a dummy sample and enclose within in it a thermocouple that is touching the bottom of the vial. This dummy sample can be frozen by direct immersion in liquid nitrogen and then allowed to thaw at room temperature. Observation of the readout generated from the thermocouple will indicate the warming rate and, typically, exposure to room temperature for more than 15–20 s will bring the outer regions of the sample close to the critical temperature of −140 °C. Warming above this level can compromise the survival of the stored sample (see Grout, Chap. 1, this Vol.).

The level of liquid nitrogen in the storage refrigerator should be monitored regularly (at least weekly and more frequently if the vessel is opened frequently during that period) and kept at the level recommended by the manufacturer. Automatic level indicator alarms and automatic filling systems are available and should be used where resources permit.

Objectives

The objectives to be met by a successful cryopreservative procedure for isolated protoplasts, cell suspension cultures and callus cultures must be:

- Preservation of optimal viability of the in vitro materials when recovered from liquid nitrogen.
- To ensure that the cell population grown from recovered material reflects the original population with regard to the characteristics required by the conserver.
- To provide the maximum genetic stability for stored materials.

3.2 Protoplast Cryopreservation by Slow Cooling

The reader is specifically referred to entries 1–13 in Table 3.1.

Equipment and materials

- Media and cryoprotectants should be prepared taking note of the warnings and guidelines presented above. The notes on the physical aspects of liquid nitrogen storage, the handling of liquid nitrogen, the sampling of populations for storage and the replication of stored cultures also apply.
- Protoplast cultures will necessarily be freshly isolated from the donor tissues to avoid the possibility of regeneration of any extracellular materials. They will be suspended in the appropriate growth medium at a density of 10^4–10^6 ml.
- Where donor material is to be acclimated to low temperatures, controlled environment conditions must be available that can provide maximum temperatures of 25 °C, with lights to simulate daytime conditions and also

Table 1. Examples of isolated protoplast, cell suspension and callus cultures that have been successfully cryopreserved using slow cooling rate techniques

1] *Zea mays*	Protoplasts	Withers 1980
2] *Daucus carota*	Protoplasts	Takeuchi et al. 1982
3] *Glycine max*	"	ibid
4] *Hordeum vulgare*	"	ibid
5] *Marchantia polymorpha*	"	ibid
6] *Triticum aestivum*	"	ibid
7] *Oryza sativa* × *Pisum sativum*	Somatic hybrid protoplasts	Bajaj 1983
8] *Oryza sativa* × *Sorghum bicolor*	"	ibid
9] *Triticum aestivum* × *Pisum sativum*	"	ibid
10] *Atropa belladona*	Protoplasts	Bajaj 1989
11] *Datura innoxia*	"	ibid
12] *Nicotiana tabacum*	"	ibid
13] *Catharanthus roseus*	Protoplasts	Gazeau et al. 1992
14] *Zea mays*	Cell suspension	Withers and King 1980
15] *Catharanthus* roseus	"	Kartha 1982
16] *Acer pseudoplatanus*	"	Withers 1985a
17] *Daucus carota*	"	ibid
18] *Hyoscyamus muticus*	"	ibid
19] *Solanum melongena*	"	ibid
20] *Sorghum bicolor*	"	ibid
21] *Musa* sp.	"	Panis et al. 1990
22] *Saccharum* sp.	"	Gnamapragasam and Vasil 1990
23] *Lavandula vera*	"	Kuriyama et al. 1990
24] *Vitis vinifera*	"	Dussert et al. 1992
25] *Asparagus officinalis*	Callus	Finkle et al. 1983
26] *Daucus carota*	"	ibid
27] *Glycine max*	"	ibid
28] *Oryza sativa*	"	ibid
29] *Phoenix dactylifera*	"	ibid
30] *Saccharum* sp.	"	ibid
31] *Triticum vulgare*	"	ibid
32] *Chamomilla recutita*	"	Cellarova et al. 1992

temperatures in the 10–0 °C range, with or without lights, to provide cold treatments.

Where in vitro cell cultures are the donor tissues, survival of isolated protoplasts following cryopreservation might be improved by acclimation of the donor material to osmotic stress. This should be attempted using sorbitol or mannitol at 6% w/v as an additive to the standard growth medium for several passages, in the first instance (see Pritchard et al. 1986).

- The standard culture medium, proven to be satisfactory for the maintenance and growth of the protoplasts, will be used for any required washing steps and post-thaw culture. Cryoprotectants are added, as required, to the complete growth medium.

- The most successful cryoprotectant systems for slow cooling-rate freezing of protoplasts have been 5% v/v DMSO + 10% carbohydrate, as additives to the standard growth medium. The solutions are:

  5% w/v DMSO 1 10%w/v glucose
  or 5% w/v DMSO + 0.25 M mannitol.

  Typically, these are formulated in standard growth medium at double the required concentration, for subsequent dilution.

- *Ampoules*. Single-use polypropylene ampoules (2 ml) with a screw cap and sealing gasket are recommended. These can be marked with a water-resistant marker pen, but scoring of identification codes lightly into the body of the tube, using a heated needle, before freezing, is recommended. Tubes can be purchased as presterilised or autoclaved in the conventional way.

- *Pipettes*. Solutions and protoplasts can be transferred using conventional sterile Pasteur and volumetric pipettes. Care must be taken to use resistant plastics when working with dimethyl sulphoxide solutions at higher concentrations.

- *Freezing and Storage Apparatus*. Slow cooling rates are most reliably achieved using purpose-built, programmable apparatus that is readily available from commercial sources. However, it is possible to construct a simple and reliable apparatus, albeit with a limited sample capacity, in the laboratory using conventional components (see Sect. 3.1 above). Storage refrigerators are widely available from commercial sources and should be liquid-nitrogen cooled, ideally with a level-monitoring safety alarm. In the author's opinion, storage in the liquid, rather than vapour, phase is preferable, and samples should be stored as multiple replicates and at different locations within the storage container. Ideally, stored samples will be replicated between duplicate storage containers.

  A number of relatively shallow wide-necked vacuum flasks are needed to hold small quantities of liquid nitrogen for various cooling procedures and for transferring frozen ampoules to the storage refrigerator. Long (20-cm), sturdy metal forceps are required.

  Liquid nitrogen is an extremely hazardous compound and should not be used without full training and use of necessary protective clothing.

- *Thawing*. Thawing is best achieved in a large (1-l) beaker of sterile water held in a thermostatted water bath at an appropriate temperature.
- *Survival Assay* (Steponkus and Lamphear 1967; Heslop-Harrison and Heslop-Harrison 1977). A freshly made stock solution of fluoroscein diacetate (0.15 w/v in absolute acetone). This solution should be stored at 4 °C. A fresh working solution is made, when required, by adding three drops of the stock solution to 10 ml of standard growth medium.

Procedures

1. Treatments to achieve acclimation of donor tissues to low growth temperatures may need to be determined empirically, with due regard to the characteristics and physiology of the species or variety. In the case of rye, for example, 1-week-old plants were moved from a 20/15 °C day/night temperature regime with a 16-h photoperiod to a 13/7 °C regime and an 11.5-h photoperiod for a further week. Thereafter, the plants were held at a constant 2 °C with a 10-h photoperiod for 4 weeks prior to protoplast isolation.
2. Cryoprotectant solutions are cooled on ice and then an equal volume added to a similarly cooled protoplast suspension, in a step-wise fashion over 30 min, followed by further incubation in the final concentration of protectant for a further 30 min before cooling and freezing.
   When designing a protocol for a new situation, a protoplast sample should be assessed for both survival and viability following cryprotectant treatment.
3. At the conclusion of the cryoprotectant incubation period 1.0 ml of protoplast suspension is transferred to a polypropylene ampoule and transferred to the freezing apparatus pre-cooled to 4 °C. A further sample should be assessed for survival/viability at this point.
4. The samples are cooled at 1 °C per min to −40 °C (see note 6). Ampoules are transferred rapidly from this temperature directly into liquid nitrogen for prolonged storage. A sample should be thawed from −35 °C (see note 6) and assayed for survival/viability. Rates between 0.8 and 2.0 °C per min have been used.
5. During storage, the level of liquid nitrogen in the storage vessel must be regularly monitored and replenished as appropriate, and good practice observed with regard to the maintenance and operation of the storage system.
6. When required, the frozen samples are removed from the storage container and thawed by immediate immersion in water at 40 °C with gentle agitation until ice is no longer visible in the sample tube.

7. The thawed samples are placed on ice and the ampoules wiped with 70% v/v ethanol before opening under sterile conditions. The protoplast suspension is carefully transferred to a larger vessel and the cryoprotectants diluted out by at least a factor of 10 by stepwise addition of the original protoplast suspension medium over 1h. The protoplasts are harvested and resuspended in an appropriate medium for subsequent regrowth or experimental use, and survival is again determined.

8. Survival is determined by placing a small sample of thawed protoplasts on a slide. Add a few drops of the diluted stain. Leave for a few minutes and then carefully cover with a coverslip and examine under the microscope, using UV light (excitation filter 355nm; barrier filter 420nm). Surviving protoplasts will show bright, clear fluorescence and should be scored for five randomly chosen fields of view at the lowest suitable magnification. A quantitative measure of survival, as a percentage, is given by:

   Number of fluorescent protoplasts/total number of protoplasts viewed $\times$ 100.

   Survival should be scored immediately post-thaw and at a suitable interval thereafter.

   Viability, however, should be scored as a measure of the number of protoplasts that either regenerate a cell wall or go on to complete cell division.

3.3 Cryopreservation of Cell Suspension Cultures by Slow Cooling

The reader is specifically referred to entries 14–24 in Table 3.1.

Successful cryopreservation of cell suspension cultures has largely relied on the following general approach:

- An extended pregrowth period, typically with a carbohydrate additive in the standard medium, which brings about both physical and chemical changes in the cultured cells which improve survival following cryopreservation. Other cryoprotectant compounds may be used as the additive.
- A cryoprotectant treatment, commonly with a mixture of protectants.
- Slow cooling, typically at 1 °C per min or less, to an intermediate temperature in the range −35 to −40 °C.

- Immersion of samples directly into liquid nitrogen from the intermediate temperature.
- Thawing by immersing vials in water at 40 °C, followed by removal of the protectants and transfer to recovery medium.

  The Withers and King protocol (see introductory paragraph, this Chap.) is the most successful variant of this procedure and is reproduced below. Details of other variations are included where appropriate.

Equipment and materials

- Media and cryoprotectants should be prepared taking note of the warnings and guidelines presented above. The notes on the physical aspects of liquid nitrogen storage, the handling of liquid nitrogen, the sampling of populations for storage and the replication of stored cultures also apply.
- Where choice exists over the selection of cultures to be frozen, or growing techniques can be applied to influence culture morphology, the investigator should look for a suspension with small cell aggregates and small, densely cytoplasmic cells. Cultures in the early exponential phase of their growth should be selected for cryopreservation.
- The pregrowth medium is the standard growth medium supplemented with 6% w/v mannitol or sorbitol. Other pregrowth additives that have been used include proline (10% w/v) and DMSO (5% v/v) (Withers 1990).
- The cryoprotectant solutions are freshly made, at twice the required final concentration, in standard liquid medium and filter sterilised.
  The preferred cryoprotectant mixture is 7.8% v/v (1 M) DMSO: 9.2% w/v (1 M) glycerol: 36% w/v (2 M) sucrose in standard liquid growth medium, which gives a final concentration of 3.9% w/v DMSO: 4.6% w/v glycerol: 18% w/v sucrose when added to the cell suspensions.

  Other successful combinations (final concentrations) that have been used include:

  5–10% v/v DMSO
  5% v/v DMSO: 5% w/v glycerol
  3.9% v/v DMSO: 4.6% w/v glycerol: 11.5% w/v proline
  10% v/v DMSO: 8% w/v glucose: 10% w/v polyethylene glycol (PEG 6000).
- Presterilised poypropylene cryogenic vials, 2 ml volume with an "O" ring seal, which are labelled using an indelible marker. An identification code can also be marked lightly into the surface of the vial using the tip of a needle.
- A controlled cooling rate apparatus capable of processing the required number of vials at rates between 0.1 and 1.0 °C per min from 0 to −35 °C, and holding them at that temperature for up to 1 h.

- Liquid nitrogen bath for plunging tubes to reach the cryogen temperature, and a liquid nitrogen vessel suitable for the long-term storage of biological materials.
- A thawing bath of sterile distilled water (to minimise the possibility of contamination) at 40 °C.
- For recovery, the standard medium solidified with agar ($7\,g\,l^{-1}$) dispensed into 5-ml petri dishes is recommended.
- Post-thaw survival can be estimated using the fluoroscein diacetate assay (see Sect. 3.2) or the Triphenyl Tetrazolium Chloride (TTC) assay (Widholme 1972). For the latter, a stock solution of 0.6% w/v TTC and 0.05% v/v Tween 80 in 0.05 M Na_2HPO_4/KH_2PO_4 buffer at ph 7.4 is required. Also needed are a solution of 95% v/v ethanol, a boiling water bath, a good light microscope and a spectrophotometer able to read at 490 nm.

Procedures

1. The selected cultures are subcultured into the pregrowth medium, and are selected for cryopreservation in the early exponential phase of growth, typically 3–6 days after the subculture.
2. A sample of suspended cells is chilled on ice and an equal volume of chilled cryoprotectant is added in three aliquots over a 20-min period. The suspension is shaken in the normal way during a 1-h incubation.
3. At the end of the incubation period, the suspension is allowed to settle, and cryoprotectant taken off to give a 30% volume of loosely packed cells.
4. One-ml aliquots are transferred to labelled polypropylene vials and then moved to the controlled-rate freezer, which is precooled to 4 °C.
5. The vials are cooled at 1 °C per min to an intermediate temperature of −35 °C, and then held at that temperature for 40 min.
6. The vials are then transferred from the freezer as rapidly as possible and plunged directly into a small Dewar of liquid nitrogen to cool them to the temperature of the cryogen. Once bubbling has ceased, the vials can be rapidly transferred to the storage vessel.
7. During storage, the level of liquid nitrogen in the storage vessel must be regularly monitored and replenished as appropriate, and good practice observed with regard to the maintenance and operation of the storage system.

8. When required, the frozen samples are removed from the storage container and thawed by immediate immersion in water at 40 °C with gentle agitation until ice is no longer visible in the sample tube.

9. The thawed samples are placed on ice and the ampoules wiped with 70% v/v ethanol before opening under sterile conditions. The cell suspension from one vial is carefully transferred to two petri dishes containing recovery medium and returned to the culture room. Any surplus liquid medium is removed from the dish 24 h after thawing.

   Alternatively, a filter paper may be placed on the surface of the recovery medium and the cells layered onto it. This paper can then be transferred to fresh medium after 1 and 24 h (Nishizawa et al. 1992; Engelmann 1992).

   At this recovery stage, the author has also to make choices concerning washing of recovered cells with fresh medium (Withers 1990) and the use of activated charcoal in the recovery medium (Kuriyama et al. 1990; Dussert et al. 1992).

10. Immediate post-thaw survival is determined either by the fluorescein diacetate assay (see Sect. 3.2) or by the TTC assay.

For the TTC assay, take approx. 100 mg fr.wt. of cells, add 3 ml of the TTC solution and infiltrate under vacuum for 10 min. Release the vacuum and incubate at 28 °C overnight. Drain away the TTC solution and wash the cells with distilled water. At this point it may be possible to look at the cells under a microscope and estimate the perentage showing the red coloration of the formazan. However, unpredictable colour development and clumping of cells makes this a difficult and generally inaccurate exercise. The washed cells are resuspended in 7 ml of 95% v/v ethanol and placed in a boiling water bath for 4 min. The samples are mixed well and then centrifuged at 1000 rpm for 5 min. The absorbance of the supernatant is measured at 490 nm and compared with the value for unfrozen control cells. An estimate of survival can be expressed as a percentage of the control value.

Viability should be scored as plating efficiency on the semi-solid medium.

3.4 Cryopreservation of Callus Cultures by Slow Cooling

Callus cultures may be treated as suspension cultures, simply by shaving off cells and aggregates from the surface of the callus and suspending them in liquid medium. The procedures in Section 3.3 will then apply. Alternatively, attempts can be made to induce a suspension by transferring callus pieces to shaken liquid medium for one or more subcultures prior to attempts at cryopreservation.

A protocol that has been successful specifically for callus tissue of a range of crop plants has been developed (Finkle et al. 1983) and is given in detail below. The reader is referred to entries 25–32 in Table 1.

Equipment and materials

- Media and cryoprotectants should be prepared taking note of the warnings and guidelines presented above (introductory paragraph, this Chap.). The notes on the physical aspects of liquid nitrogen storage, the handling of liquid nitrogen, the sampling of populations for storage and the replication of stored cultures also apply.

- Where choice exists over the selection of cultures to be frozen, or growing techniques can be applied to influence culture morphology, the investigator should look for callus masses with small cell aggregates and small, densely cytoplasmic cells. The callus should be selected at the mid-point of an optimum growth cycle, typically 1–2 weeks after transfer to fresh medium.

- The cryoprotectant solutions are freshly made, at twice the required final concentration, in standard liquid medium and filter sterilised. Autoclaving is not appropriate, given the nature and concentration of the compounds concerned.

  The preferred cryoprotectant mixture is 20% v/v DMSO: 20% w/v polyethylene glycol: 16% w/v sucrose in standard liquid growth medium, which gives a final concentration of 10% w/v DMSO: 10% w/v polyethgylene glycol: 8% w/v sucrose when added to the cell suspensions.

- Presterilised poypropylene cryogenic vials, 2 ml volume with an "O" ring seal, which are labelled using an indelible marker. An identification code can also be marked, lightly, into the surface of the vial using the tip of a needle.

- A controlled cooling rate apparatus capable of processing the required number of vials at 1.0 °C per min from 0 to −35 °C (see above and Withers and King 1980; Withers 1990).

- Liquid nitrogen bath for plunging tubes to reach the cryogen temperature, and a liquid nitrogen vessel suitable for the long-term storage of biological materials.

- A thawing bath of sterile distilled water (to minimise the possibility of contamination) at 40 °C.

- For recovery, the standard liquid medium is required for washing and the same medium, solidified with agar ($7\,g\,l^{-1}$), dispensed into 5-ml petri dishes is required for recovery growth.

Procedures

1. The selected callus masses (max. 5 mm cube) are immersed in liquid standard culture medium and chilled on ice.
2. An equal volume of chilled cryoprotectant is added in three aliquots over the first 20 min of a 1-h incubation.
3. At the end of the incubation period, the suspension is allowed to settle, and cryoprotectant taken off to give a 30% volume of loosely packed cells.
4. Two or three callus pieces are transferred to a labelled polypropylene vial and then moved to the freezer, which is precooled to 4 °C.
5. The vials are cooled at 1 °C per min to an intermediate temperature of −35 °C.
6. The vials are then transferred from the freezer as rapidly as possible and plunged directly into a small Dewar of liquid nitrogen to cool them to the temperature of the cryogen. Once bubbling has ceased, the vials can be rapidly transferred to the storage vessel.
7. During storage, the level of liquid nitrogen in the storage vessel must be regularly monitored and replenished as appropriate, and good practice observed with regard to the maintenance and operation of the storage system.
8. When required, the frozen samples are removed from the storage container and thawed by immediate immersion in water at 40 °C with gentle agitation until ice is no longer visible in the sample tube.
9. The thawed samples are placed on ice and the ampoules wiped with 70% v/v ethanol before opening under sterile conditions. The callus pieces are washed with fresh medium at room temperature and carefully transferred to the petri dishes containing recovery medium and returned to the culture room. Any surplus liquid medium is removed from the dish 24 h after thawing.
10. Immediate post-thaw survival may be estimated using the TTC assay (see Sect. 3.3), and viability by recording the number of callus pieces that sustain an increase in size in the recovery period. A lag of several weeks is not uncommon before an obvious increase in the size of the thawed callus pieces is observed.

References

Bajaj YPS (1983) Survival of somatic hybrid protoplasts of wheat × pea and rice × pea subjected to −196 °C. Indian J Exp Biol 21:120–122

Bajaj YPS (1988) Regeneration of plants from frozen (−196 °C) protoplasts of *Atropa belladona* L., *Datura innoxia* Mill. and *Nicotiana tabacum* L. Indian J Exp Biol 26:289–292

Bajaj YPS (1989) Cryopreservation of plant protoplasts. In: Bajaj YPS (ed) Biotechnology in agriculture and forestry, vol 8: Plant protoplasts and genetic engineering. Springer, Berlin Heidelberg New York, pp 97–106

Cellarova E, Cernicka T, Vranova E, Brutovska R, Lapar M (1992) Viability of *Chamomilla recutita* L. cells after cryopreservation. Cryo Lett 13:37–42

Cocking EC (1972) Plant cell protoplasts – isolation and development. Annu Rev Plant Physiol 23:29–50

Dodds JH (1991) In vitro methods for conservation of plant genetic resources. Chapman & Hall, London

Dussert S, Mauro MC, Deloire A, Hamon S, Engelmann F (1991) Cryopreservation of grape embryogenic cell suspensions 1. Influence of pretreatment, freezing and thawing conditions. Cryo Lett 12:287–298

Dussert S, Mauro MC, Engelmann F (1992) Cryopreservation of grape embryogenic cell suspensions 2. Influence of post-thaw culture conditions and application to different strains. Cryo Lett 13:15–22

Engelmann F (1992) Effect of freezing in liquid nitrogen on the properties of a soybean (*Glycine max* L. var. acme) callus strain used as a bioassay for cytokinin activity. Cryo Lett 13:331–336

Finkle BJ, Ulrich JM, Schaeffer GW, Sharpe F Jr (1983) Cryopreservation of rice cells. In: Inst of Genetics Academis Sinica/Int Rice Res Inst (eds) Cell and tissue culture techniques for cereal crop improvement. Gordon & Breach, New York, pp 343–369

Gazeau CM, Blanchon C, Dereuddre J (1992) Freeze-preservation of *Catharanthus* protoplasts at liquid nitrogen temperature. Comparison with cells. Cryo Lett 13:149–158

Gnamapragasam S, Vasil K (1990) Plant regeneration from a cryopreserveed embryogenic cell suspension of a commercial sugarcane hybrid (*Saccharum* sp.). Plant Cell Rep 9:419–423

Grout BWW, Morris GJ (1987) The effects of low temperatures on biological systems. Edward Arnold, London

Heslop-Harrison J, Heslop-Harrison Y (1977) Evaluation of pollen viability by enzymatically induced fluorescence: intracellular hydrolysis of fluorescein diacetate. Stain Technol 45:115–121

Kartha KK (1982) Cryopreservation of periwinkle, *Catharanthus roseus*, cells cultured in vitro. Plant Cell Rep 1:135–138

Kartha K (ed) (1985) Cryopreservation of plant cells and organs. CRC Press, Boca Raton

Kuriyama A, Watanabe K, Ueno S, Mitsuda H (1990) Effect of post-thaw treatment on the viability of cryopreserved *Lavandula vera* cells. Cryo Lett 11:171–178

Nishizawa S, Sakai A, Amano Y, Matsuzawa T (1992) Cryopreservation of asparagus (*Asparagus officinalis*) embryogenic cells and subsequent plant regeneration by a simple freezing method. Cryo Lett 13:379–388

Panis BJ, Withers LA, De Langhe EAL (1990) Cryopreservation of *Musa* suspension culutres and subsequent regeneration of plants. Cryo Lett 11:337–350

Power JB, Davey M (1990) Protoplasts of higher and lower plants. In: Pollard JW, Walker JM (eds) Methods in molecular biology, vol 6: Plant cell and tissue culture. Humana Press, New Jersey, pp 237–259

Pritchard HW, Grout BWW, Short KC (1986) Osmotic stress as a pregrowth procedure for cryopreservation 3. Cryobiology of sycamore and soybean cell suspensions. Ann Bot 57:379–388

Steponkus PL (1985) Cryobiology of isolated protoplasts – applications to plant cell cryopreservation. In: Karhta K (ed) Cryopreservation of plant cells and organs. CRC Press, Boca Raton, pp 49–60

Steponkus PL, Lamphear FO (1967) Refinement of the triphenyl tetrazolium chloride method of determining cold injury. Plant Physiol 42:1423–1426

Takeuchi M, Matushima H, Sugawara Y (1982) Totipotency and viability of protoplasts after long-term freeze preservation. In: Fujiwara A (ed) Plant tissue culture 1982. Japanese Association for Plant Tissue Culture, Tokyo, pp 797–798

Widholme JM (1972) The use of fluorescein diacetate for determining viability of cultured plant cells. Stain Technol 47:189–194

Withers LA (1980) Preservation of germplasm. In: Vasil IK (ed) International review of cytology. Supplement 11B Perspectives in plant cell and tissue culture. Academic Press, New York, pp 101–136

Withers LA (1985a) Cryopreservation of cultured plant cells and protoplasts. In: Kartha KK (ed) Cryopreservation of plant cells and organs. CRC Press, Boca Raton, pp 244–267

Withers LA (1985b) Cryopreservation and storage of germplasm. In: Dixon RA (ed) Plant cell cultures – a practical approach. IRL Press, Oxford, pp 169–192

Withers LA (1987) The low-temperature preservation of plant cell, tissue and organ cultures and seed for genetic conservation and improved agricultural practice. In: Grout BWW, Morris JG (eds) The effects of low temperatures on biological systems. Edward Arnold, London, pp 389–409

Withers LA (1990) Cryopreservation of plant cells. In: Pollard JW, Walker JM (eds) Methods in molecular biology, vol 6. Plant cell and tissue culture. The Humana Press, New Jersey, pp 39–48

Withers LA, King PJ (1980) A simple freezing unit and cryopreservation method for plant cell suspensions. Cryo Lett 1:213–220

4 Cryopreservation of Excised Meristems, Shoot Tips and Somatic Embryos

B.W.W. Grout

The cryopreservation of in vitro cultures of excised meristems and somatic embryos is not an area that has been extensively researched and has, therefore, a relatively retricted experimental literature. Whilst there is no single, definitive protocol which can be shown to be suitable for anything like a comprehensive range of plant species, there are trends and protocols with promise that have been successful for a number of species, and it is these that are reported below.

In the cryopreservation literature the terms "meristem" and "shoot tip" are often used as if they were interchangeable, and commonly without any explanation of the size of the explant concerned. Excised meristems are usually of the order of 0.4–1.0 mm from apex to base (Kartha 1985; Towill 1991) and comprise the apical dome and several of the very smallest leaf primordia, whereas shoot tips can be assumed to be somewhat larger and with more, better developed leaf primordia. In the discussion below we have, where required, used the original term given by the authors.

Excised meristems are poor candidates for preservation by slowing, but not stopping, growth and development (see Grout, Chap. 2, this Vol.), as it is essential for their survival that they move at least some way along their developmental pathway, forming a larger structure and diminishing the structural and functional effects of excision. Development of an in vitro meristem culture may take one of two forms, and the most desirable, from the conservation point of view, is direct outgrowth of the meristem to form a plantlet. The second, and less desirable, pattern results in callusing from the excision wound and expanding leaf primordia. This early callus very frequently develops adventitious meristems, and thereafter plantlets. The disadvantage is that the intervention of a callus phase greatly enhances the possibility of somaclonal variation in any derived population of plantlets (Grout, Chap. 1, this Vol.; Phillips et al. 1990).

The preferred option for the extended storage of meristems is cryopreservation, and yet in this respect, the excised meristem poses particular problems due to its mass and multicellular diversity. For example, the compact structure of the meristem will limit the extracellular space available for water during cryodehydration (see Grout, Chap. 1, this Vol.). The mass of the meristem will also influence the diffusion of cryoprotectants into the cells of the explant, making equilibration of each individual cell with the external levels more difficult. Structural diversity also presents difficulties as, for example, the degree of vacuolation in cells behind the meristematic dome can increase from 15 to 80% of cell volume over a distance of no more than five cell diameters. Attempting to devise a single protocol that is effective in preserving such a diverse range of cell types, and thus preserving the meristem in its entirety, is very difficult. As increasingly less of the full complement of cells in the explant survives the cryopreservative procedure, so the likelihood of disorganised growth after thawing is increased. Similar problems will occur with somatic embryos when they have differentiated meristematic regions. Indeed, it is probably a very exceptional circumstance when material is recovered from cryopreservation without cellular damage, and as experience with cryopreserved plant culture systems has accumulated, the concept of a phase of repair following thawing, as part of an optimal protocol, has become accepted (Grout 1991). Both environmental and chemical manipulations and conditions may have an effect on the regrowth process and should be considered as a part of the empirical scheme that may be needed to optimise a protocol. For example, high levels of post-thaw light may cause photobleaching with adverse effects on viability and so low light regimes are to be recommended during early regrowth (Benson et al. 1989). The duration of the in vitro period of the cultures before cryopreservation may also be significant with regard to final recovery. In general, the successful regrowth of thawed organs is dependent upon the presence of exogenous plant growth regulators in the recovery medium, but they may not be the same, either quantitatively or qualitatively, as those required for the growth of material that has not been cryopreserved (Grout et al. 1978; Towill 1981; Grout 1991).

The quantity of reported experimental work that has been carried out on the cryopreservation of somatic embryos is even more limited than for excised meristems. Consequently, there are several approaches which may eventually prove to be of broad application but there is little evidence, to date, as to which approach will be the most successful. The investigator faced with the challenge of cryoconserving somatic embryos of species not previously studied must, therefore, expect to be involved in experimental work to produce a successful protocol. The information given below is intended to provide a first procedure and, if this is unsuccessful, to be a useful stepping off point for further study.

Note For the manipulations described below, from in vitro pretreatment onwards, sterile procedures are required. Autoclaved media and reagents, sterile glassware, vials and other disposables should be used. All manipulations should be carried out at a sterile-air work station with due regard to the normal practices involved in plant tissue and organ culture.

Objectives

The objectives to be met by a successful cryopreservative procedure for excised shoot tips, meristems or somatic embryos must be:

- Preservation of optimum viability of the in vitro materials when recovered from liquid nitrogen.
- To ensure that the plant regeneration from viable explants following cryopreservation is by outgrowth of original frozen/thawed meristems, and not from adventitious growth.
- To provide the maximum genetic stability for stored explants.

4.1 Meristem/Shoot-Tip Cryopreservation by Slow Cooling

The reader is specifically referred to entries 2, 8, 11–17 and 19 in Table 1.

Whilst there is no single procedure in this category that is of universal application, there is a general approach that appears to be successful. This is:

General approach

- Where the species concerned has some degree of cold-hardening capability, there is benefit in pregrowth of donor plants or in vitro cultures at low temperatures to bring about the cellular changes associated with acclimation to low temperature (see Grout, Chap. 1, this Vol.).
- Pretreatment of the excised meristem to overcome the damage associated with excision and, in some cases, to provide prolonged exposure to relatively low levels of compounds with cryoprotectant properties.
- Treatment with cryoprotectant in which the explants are exposed to the final levels of the compounds in which they will be frozen.
- Cooling at a rate of 1 °C per min or less to an intermediate temperature, (typically −40 °C), followed by plunging into liquid nitrogen.
- Thawing by immersion in water at 40 °C.

Equipment and materials

- Media and cryoprotectants should be prepared taking note of the warnings and guidelines previously presented (see Grout, Chap. 3, this Vol.). The notes on the physical aspects of liquid nitrogen storage, the handling of liquid nitrogen, the sampling of populations for storage and the replication of stored cultures also apply.
- Where pregrowth leading to cold hardening of donor plants or in vitro cultures is to be attempted, then appropriate controlled environmental

Table 1. Examples of isolated meristem/shoot tip cultures and somatic embryos exhibiting plant regeneration after cryopreservation

1] *Arachis hypogaea*	Apical meristems	Bajaj 1979
2] *Fragaria* × *ananassa*	Apical meristems	Kartha et al. 1980
3] *Solanum tuberosum*	Apical meristems	Bajaj 1981
4] *Manihot esculenta*	Apical meristems	Bajaj 1983
5] *Pyrus serotina*	Shoot tips	Moriguchi et al. 1985
6] *Rubus* sp.	Apical meristems	Reed and Lagerstedt 1987
7] *Dianthus caryophyllus*	Shoot tips	Dereuddre et al. 1988
8] *Beta vulgaris*	Apical meristems	Braun 1988
9] *Haplopappus gracillis*	Shoot primordia	Taniguchi et al. 1988
10] *Mentha* sp.	Shoot tips	Towill 1991
11] *Brassica napus*	Shoot tips	Withers et al. 1988
12] *Vaccinium* sp.	Apical meristems	Reed 1989
13] *Solanum tuberosum*	Shoot tips	Harding et al. 1991
14] *Trifolium repens*	Apical meristems	Yamada et al. 1991
15] *Chrysanthemum morifolium*	Shoot tips	Fukai et al. 1991
16] *Phoenix dactylifera*	Apical meristems	Bagniol & Engelmann 1991
17] *Prunus salicina* × *spinosa*	Shoot tips	Brison et al. 1992
18] *Ribes aureum*	Apical meristems	Reed 1992
19] *Cichorium intypbus*	Shoot tips	Demeulemeester et al. 1993
20] *Dianthus caryophyllus*	Shoot tips	Seibert and Wetherbee 1977
21] *Solanum goniocalyx*	Shoot tips	Grout and Henshaw 1978
22] *Dianthus caryophyllus*	Shoot apices	Uemuru and Sakai 1980
23] *Asparagus officinalis*	Shoot tips	Kumu et al. 1983
24] *Digitalis lanata*	Shoot tips	Diettrich et al. 1987
25] *Brassica napus*	Shoot tips	Withers et al. 1988
26] *Solanum tuberosum*	Shoot tips	Benson et al. 1989
27] *Daucus carota*	Somatic embryos	Withers 1979
28] *Elaeis guineensis*	Somatic embryos	Engelmann et al. 1985
29] *Citrus sinensis*	Somatic embryos	Marin et al. 1988
30] *Elaeis guineensis*	Somatic embryos	Engelmann 1990
31] *Daucus carota*	Somatic embryos	Dereuddre et al. 1991
32] *Cucumis melo*	Somatic embryos	Shimonishi et al. 1991

conditions must be available. These should be able to provide low temperatures in the 0–10 °C range, with or without lights, and maximum temperatures of 25 °C with lights to simulate daytime conditions.

- Cryopreservation is best carried out in polypropylene vials (1–2.5 ml volume) designed for liquid nitrogen storage. These should be labelled as described in Chapter 3, this Volume.
- A range of prepared media is required:
  - Standard nutrient medium;
  - pretreatment medium which will be one of two types. Either standard medium plus sucrose (0.1–0.5 M) *or* standard medium plus DMSO

(5% w/v). This may be supplemented with glucose [5% w/v – Table 4.1 (14)] or proline (2% w/v – Table 4.1 (17)].
- recovery medium which will be based upon the standard medium, but may have an altered composition with regard to carbohydrate levels and the level and types of plant growth regulators (see introductory paragraph, this Chap.). Any alterations must be determined empirically.

- *Cryoprotectant solutions.* These are based on standard medium with the addition of various cryoprotectant compounds and fall into three categories:
  - Standard medium with DMSO (5–15% w/v);
  - standard medium with DMSO (5–15% w/v) with a carbohydrate additive, e.g., 10% w/v glucose, sucrose or sorbitol 5–15% w/v [see Table 1, (14, 15, 16, 8, resp.)];
  - standard medium with multiple cryoprotectant additives, e.g., glycerol : ethylene glycol:DMSO:sucrose at 30:15:10:4% w/v or polyethylene glycol:glucose:DMSO at 10:10:10% w/v [see Table 1, (17, 12)].
- Controlled cooling rate apparatus capable of processing the required number of vials at rates between 0.1 and 1 °C per min to −40 °C, and holding them at that temperature for up to 1 h.
- Liquid nitrogen bath for plunging vials to achieve cryogen temperature, and a liquid nitrogen vessel suitable for the long-term storage of biological materials.
- A thawing bath of sterile distilled water at 40 °C.

Procedures

1. If the donor source of the meristem/shoot-tip explants is likely to acclimate to cold, then a suitable hardening regime has to be devised. For an entire plant raised either in the field or a glasshouse, it is possible to let the material harden in a natural winter and excise the meristems at, physiologically, the peak of hardiness. Alternatively, a typical mid-severe natural hardening regime can be simulated in an environment chamber. Where the donor material is an in vitro plantlet, a regime of temperate days (22 °C) and cool nights (−1 to 4 °C) may be applied for a period of 7–28 days. Optimal conditions will need to be determined empirically, and leaflet frost-hardiness tests may be of value (Quamme 1978).
2. Following aseptic excision of the meristem or shoot tip, a pretreatment before cryoprotectant exposure and freezing is generally of benefit. At

its simplest, this will be 24–48 h on standard medium, and sucrose supplements have been employed from 0.1–0.5 M. Most commonly used is standard medium plus 5% w/v DMSO for 24–48 h. On occasions this has been supplemented with proline (2% w/v) or glucose (5% w/v at 4 °C).

3. A suitable cryoprotectant treatment may be chosen on the grounds of similarity between the subject of the study and one of the examples cited in Table 1. In the absence of any such suitable indicator, possible cryoprotectants must be screened for toxicity without freezing, using concentration and time as the variables for the investigation.

Cryoprotectant exposure will be for up to 1 h, and the longer times are especially applicable if the cryoprotectant incubation is carried out on ice, which may moderate toxic effects to some extent.

Even under routine conditions, and where the quantity of material permits, there is merit in taking a sample of meristems/shoot tips for regrowth after the cryoprotectant solution.

4. Cooling is controlled from 0 °C to an intermediate temperature of −35 or −40 °C, at a rate between 0.1 and 1.0 °C per min, with rates between 0.5 and 1.0 °C per min being preferred.

Induced seeding of ice, typically at temperatures between −5 and −10 °C, during cooling is not a procedural step commonly reported in the published literature, but it is to be recommended as it produces greater control over cooling rate and a high degree of uniformity over processing and the thermal histories of individual vials (see Grout, Chap. 1, this Vol.).

5. At the intermediate temperature, a period (up to 30 min) can be inserted into the procedure to prolong cryodehydration. In any investigation to optimise the cryopreservation protocol, the regrowth results of a sample thawed from the intermediate temperature will be particularly valuable.

From the intermediate temperature, individual vials are plunged directly into liquid nitrogen in a small Dewar container to achieve reduction to −196 °C. Thereafter the vials are very rapidly transferred to the cryogenic storage system, taking care to avoid more than a few seconds of rewarming at any stage in the transfer.

6. During storage, the level of liquid nitrogen in the storage vessel must be regularly monitored and replenished as appropriate, and good practice observed with regard to the maintenance and operation of the storage system (see Grout, Chap. 2, this Vol.).

7. Vials are thawed by immersing them, individually, in sterile distilled water warmed to 40 °C. The vials are agitated during the thawing period and removed from the warming bath when ice crystals are no longer visible in the contained solutions.
8. Once thawed, the tissues are transferred to the appropriate regrowth medium (see introductory paragraph, this Chap.)
9. There is no immediate viability test that can be reliably employed to assess the success of cryopreservation of meristems and shoot tips immediately after thawing. Regrowth should be monitored using a high-powered hand lens as well as conventional light and scanning electron microscopy to determine the pattern of plant regeneration. Acceptable regrowth is from the original meristems of the frozen/thawed explants and not adventitious in origin.

4.2 Meristem/Shoot-Tip Cryopreservation by Rapid Cooling

The reader is specifically referred to entries 20–26 in Table 1.

Equipment and materials

- Media and cryoprotectants should be prepared taking note of the warnings and guidelines presented in Chapter 3. The notes on the physical aspects of liquid nitrogen storage, the handling of liquid nitrogen, the sampling of populations for storage and the replication of stored cultures also apply.
- The choices of media and cryoprotectants are essentially those presented in Section 4.1.
- Rapid freezing requires either polypropylene cryopreservation vials or a physical carrier such as a hypodermic needle mounted on a 1-ml syringe barrel, or similar. The key feature is that the metal needle or other carrier should be readily detachable from the holder.
- All other equipment is as in Section 4.1.

Procedures

1. Donor material is pregrown and cold-hardened as appropriate.
2. Shoot tips or meristems are excised from the donor material and pregrown as required.
3. Following incubation in cryoprotectant the explants can be rapidly cooled in one of two ways:

4A. One or two meristems are lifted from the cryoprotectant solution onto the tip of a hypodermic needle or similar metal carrier. The carrier is plunged into directly liquid nitrogen through a minimum depth of 7 cm.

5A. When the nitrogen has ceased bubbling, the carrier can be detached from its holder and transferred to a propylene cryopreservation vial, very loosely plugged with cotton wool as a closure, for subsequent storage.

6A. During storage, the level of liquid nitrogen in the storage vessel must be regularly monitored and replenished as appropriate, and good practice observed with regard to the maintenance and operation of the storage system (see Chap. 2, this Vol.).

7A. To thaw, the vial is taken from storage and immersed in a shallow dish of liquid nitrogen. After unplugging, the carriers are removed and immersed in liquid medium at 25 °C. The thawed shoot tips will float away from the carrier to be collected for subsequent regrowth.
Continue with step 8.
or:

4B. After cryoprotection, the meristems are transferred to a polypropylene vial with 0.5–1.0 ml of the cryoprotectant medium.

5B. The vials are frozen by direct immersion in liquid nitrogen and are transferred into storage after bubbling of the nitrogen has ceased.

6B. During storage, the level of liquid nitrogen in the storage vessel must be regularly monitored and replenished as appropriate, and good practice observed with regard to the maintenance and operation of the storage system (see Grout, Chap. 3, this Vol.).

7B. To thaw, the vial is taken from storage and immersed in sterile distilled water warmed to 40 °C. The vial is agitated during the thawing period and removed from the warming bath when ice crystals are no longer visible in the contained solutions.

8. Once thawed, the meristems are transferred to the appropriate regrowth medium.

9. Regrowth should be monitored as in Section 4.1.

4.3 Cryopreservation of Somatic Embryos by Dry Freezing

The reader is specifically referred to entries 27 and 33 in Table 1.

Equipment and materials

- Media and cryoprotectants should be prepared taking note of the warnings and guidelines presented in Chapter 3, this Volume. The notes on the physical aspects of liquid nitrogen storage, the handling of liquid nitrogen, the sampling of populations for storage and the replication of stored cultures also apply.
- A cryoprotectant medium based on the standard growth medium with DMSO as an additive. A level of 10% w/v is suggested for initial studies.
- Small, autoclaved envelopes (5 × 5 cm) made from domestic aluminium foil folded several times to ensure a seal at the edges. An indelible system of labelling must be provided on the outside of the envelope.
  As a security measure, coloured, autoclavable beads or other suitable markers can be placed inside the envelope to confirm identification.
- Container of ice.
- Controlled cooling rate apparatus to produce cooling rates of 0.5–1.0 °C per min from 0 to −40 °C.
- Liquid nitrogen bath for plunging envelopes to achieve cryogen temperature, and a liquid nitrogen vessel suitable for the long-term storage of biological materials.
- During storage, the level of liquid nitrogen in the storage vessel must be regularly monitored and replenished as appropriate, and good practice observed with regard to the maintenance and operation of the storage system (see Chap. 2, this Vol.).
- Regrowth medium which may be standard liquid medium, standard semi-solid medium or standard semi-solid medium with 1% w/v activated charcoal.

Procedures

1. Mature somatic embryos are harvested from standard liquid culture conditions.
2. Incubate embryos in standard medium plus the selected DMSO additive, on ice, for 1 h.
3. Remove embryos from the cryoprotectant medium and blot dry on sterile filter paper.
4. Place five to six embryos in a foil envelope and seal securely by folding over the open edge several times. Ensure that each envelope is adequately labelled.
5. Cool the envelope at 1 °C per min from 0 to −40 °C and then, avoiding any more than a few secods in which warming might occur, plunge the

envelope directly into liquid nitrogen. Once bubbling of the nitrogen has ceased, the envelope can be transferred to the storage container.

6. To thaw, the envelope is taken from the storage container and left at room temperature for 15 min.

7. The thawed embryos are transferred to

- Standard liquid medium which is changed after 1 and 12 h or
- semi-solid medium with/without activated charcoal.

8. The surviving embryos should be monitored to see if subsequent plant regeneration is from the original embryo or secondary embryogenesis.

4.4 Cryopreservation of Somatic Embryos by Rapid Cooling

This protocol is particularly appropriate for early-stage somatic embryos and also for embryogenic regions developed in callus tissue growing on semi-solid medium.
The reader is specifically referred to entries 29 and 30 in Table 1.

Equipment and materials

- Media and cryoprotectants should be prepared taking note of the warnings and guidelines presented in Chapter 3, this Volume. The notes on the physical aspects of liquid nitrogen storage, the handling of liquid nitrogen, the sampling of populations for storage and the replication of stored cultures also apply.
- A semi-solid pregrowth medium which is the appropriate standard medium with 0.75 M sucrose as an additive.
- Polypropylene cryogenic vials.
- Liquid nitrogen bath for plunging vials to achieve cryogen temperature, and a suitable liquid nitrogen vessel suitable for the long-term storage of biological materials.
- A thawing bath of sterile distilled water at 40 °C.
- A regrowth medium based on the standard semi-solid formulation with 0.1 M sucrose as an additive.

Procedures

1. The culture of immature embryos or embryogenic callus is transferred to standard semi-solid medium supplemented with 0.75 M sucrose for a pretreatment lasting 7 days.

2. Where appropriate, embryogenic masses are dissected from the surrounding undifferentiated callus mass.

3. Immature embryos or embryogenic tissue masses are transferred to polypropylene cryogenic vials. A total tissue volume of 0.5 ml per vial should not be exceeded.

4. Capped vials are immersed directly in liquid nitrogen and then, once bubbling of the nitrogen has ceased, are rapidly transferred to the storage vessel.

5. During storage, the level of liquid nitrogen in the storage vessel must be regularly monitored and replenished as appropriate, and good practice observed with regard to the maintenance and operation of the storage system (see Grout, Chap. 3, this Vol.).

6. Vials are thawed by immersion in water at 40 °C for 1.5 min.

7. Thawed embryos and embryogenic masses are transferred to regrowth medium for 2 weeks and to standard medium thereafter.

4.5 Cryopreservation of Somatic Embryos Following Dessication

The reader is specifically referrred to entries 32 and 33 in Table 1

The generalised procedure appears to exploit an ability to survive desiccation that is expressed in the somatic embryo in the same manner as in the zygotic embryo. The technique has been successful both with and without alginate encapsulation.

Equipment and materials

- Media and cryoprotectants should be prepared taking note of the warnings and guidelines presented in Chapter 3, this volume. The notes on the physical aspects of liquid nitrogen storage, the handling of liquid nitrogen, the sampling of populations for storage and the replication of stored cultures also apply.

Without alginate encapsulation

- A pretreatment medium based on the standard medium with abscisic acid ($10\,mg\,l^{-1}$) as an additive.

- Standard medium for regrowth.

With alginate encapsulation

- A pretreatment medium with 0.3 M sucrose is used as an additive.
- For encapsulation: calcium-free culture medium; 6% Na-alginate solution; culture medium supplemented with 0.1 M $CaCl_2$.
- A semi-solid medium (0.6% w/v agar) for regrowth with a sucrose level of 0.015 M.
- A facility to desiccate embryos under sterile conditions. The simplest system is to use the airflow in the sterile work station and to monitor water loss gravimetrically.
- Polypropylene cryogenic vials or aluminium foil envelopes (see Sect. 4.3)
- Controlled cooling rate apparatus to produce cooling rates of 0.5–1.0 °C per min from 0 to −40 °C.
- Liquid nitrogen bath for plunging envelopes to achieve cryogen temperature, and a suitable liquid nitrogen vessel suitable for the long-term storage of biological materials.

Procedures

Without alginate encapsulation

1. Transfer embryos to the pregrowth medium for 3 days.
2. Desiccate to a water content of ca. 12% of fresh weight.
3. Transfer embryos to a polypropylene cryogenic vial or a foil envelope and freeze by direct immersion in liquid nitrogen.
4. Thaw at room temperature for 15 min and then transfer embryos back to standard medium for regrowth.
5. The surviving embryos should be monitored to see if subsequent plant regeneration is from the original embryo or secondary embryogenesis.

With alginate encapsulation

1. Mature somatic embryos are harvested from liquid culture and incorporated into alginate beads (see below and Redenbaugh et al. 1988).
2. Suspend the embryos in calcium-free culture medium and mix with an equal volume of 6% w/v Na-alginate solution to achieve a final density of six to eight embryos per ml.

3. Beads are made by dropping this mixture into culture medium supplmented with calcium chloride. Beads of approx. 4 mm in diameter and one or two embryos are collected.
4. The selected beads are pretreated by transfer to culture medium with 0.3 M sucrose for 18 h.
5. The beads are dried in the airflow of the sterile work station for 4 h and then transferred into polypropylene cryogenic vials (four beads per vial).
6. The vials are cooled by direct immersion in liquid nitrogen or at 0.5 °C per min from 0 to −40 °C before immersion in liquid nitrogen. Subsequently, the vials are transferred to a suitable storage vessel.
7. The frozen "synthetic seeds" are thawed at room temperature and placed directly onto the regrowth medium for subsequent "germination". This material should be examined for direct, or secondary, regrowth.

References

Ammirato PV (1983) Embryogenesis. In: Evans DA, Sharp WR, Ammirato PV, Yamada Y (eds) Handbook of plant cell culture, vol 1: Techniques of propagation and breeding. Macmillan, New York, pp 82–123

Bagniol S, Engelmann F (1991) Effects of pregrowth and freezing conditions on the resistance of meristems of date palm to freezing in liquid nitrogen. Cryo Lett 12:279–286

Bajaj YPS (1979) Freeze preservation of meristems of *Arachis hypogea* and *Cicer arietinum* Indian J Exp Biol 17:1407–1407

Bajaj YPS (1981) Regeneration of plants from potato meristems freeze-preserved for 24 months. Euphytica 30:141–145

Bajaj YPS (1983) *Cassava* plants from meristem cultures freeze-preserved for three years. Field Crops Res 7:161–167

Benson EE, Harding K, Smith H (1989) The effects of pre- and post-freeze light on the recovery of cryopreserved shoot tips of *Solanum tuberosum*. Cryo Lett 10:323–344

Braun A (1988) Cryopreservation of sugar beet germplasm. Plant Cell Tissue Organ Cult 14:161–168

Brison M, Paulus V, deBoucard MT, Dosba F (1992) Cryopreservation of walnut and plum shoot tips. Cryobiology 29:738

Demeulmeester MAC, Vandenbussche B, DeProft MP (1993) Regeneration of chicory plants from cryopreserved in vitro shoot tips. Cryo Lett 14:57–64

Dereuddre J, Fabre J, Bassaglia C (1988) Resistance to freezing in liquid nitrogen of carnation (*Dianthus caryophyllus* L. var. Eolo) apical and axillary shoot tips excised from different aged in vivo plants. Plant Cell Rep 7:170–173

Dereuddre J, Blandin S, Hassen N (1991) Resistance of alginate-coated somatic embryos of carrot (*Daucus carota* L.) to desiccation and freezing in liquid nitrogen. 1. Effects of preculture. Cryo Lett 12:125–134

Diettrich B, Wolf T, Bormann A, Popov AS, Butenko RG, Luckner M (1987) Cryopreservation of *Digitalis lanata* shoot tips. Plant Med 53:359–363

Engelmann F (1990) Cryoconservation des embryos somatiques de palmier à huile. Bull Soc Bot Fr Actual Bot 137:93–98

Engelmann F, Duval Y, Dereuddre J (1985) Survival and proliferation of oil palm (*Elaeis guineensis* Jacq.) somatic embryos after freezing to liquid nitrogen. CR Acad Sci 301:111–116

Fukai S, Goi M, Tanaka M (1991) Cryopreservation of shoot tips of *Chrysanthemum morifolium* and related species native to Japan. Euphytica 54:201–204

Grout BWW (1991) Cryopreservation of plant cells and organs. In: Zakri AH, Normah MN, Senawi MT, Abdul Karim AG (eds) Conservation of plant genetic resources through in vitro methods. FRIM/MNCPGR, Kuala Lumpur, pp 43–55

Grout BWW, Henshaw GG (1987) Freeze-preservation of potato shoot-tip cultures. Ann Bot 42:1227–1229

Grout BWW, Westcott RJ, Henshaw GG (1978) Survival of shoot meristems of tomato seedlings frozen in liquid nitrogen. Cryobiology 15:478–483

Harding K, Benson EE, Smith H (1991) The effects of pre-freeze in vitro culture period on the recovery of cryopreserved shoot tips of *Solanum tuberosum*. Cryo Lett 12:17–22

Kartha KK (1985) Meristem culture and germplasm preservation. In: Kartha KK (ed) Cryopreservation of plant cells and organs. CRC Press, Boca Raton, pp 115–134

Kartha KK, Leung NL, Pahl K (1980) Cryopreservation of strawberry meristems and mass propagation of plantlets. J Am Soc Hortic Sci 105:481–484

Kumu Y, Harada T, Yakuwa T (1983) Development of a whole plant from a shoot tip of *Asparagus officinalis* frozen down to −196 °C. J Fac Agric Hakkaido Univ 61:285–294

Marin ML, Duran-Vila N (1988) Survival of somatic embryos and recovery of plants of sweet orange (*Citrus sinensis* L. Osb.) after immersion in liquid nitrogen. Plant Cell Tissue Organ Cult 14:51–57

Moriguchi T, Akihama T, Kozai I (1985) Freeze preservation of dormant pear shoot apices. Jpn J Breed 35:196–199

Phillips RL, Kaeppler SM, Peschke VM (1990) Do we understand somaclonal variation? In: Nijkamp HJJ, van der Plas LHW, van Aartrijk J (eds) Progress in plant cellular and molecular biology. Kluwer Academic Publishers, Dordrecht, pp 131–141

Quamme HA (1978) Breeding and selecting temperate fruit crops for cold hardiness. In: Li PH, Sakai A (eds) Plant cold hardiness and freezing stress. Academic Press, London, pp 313–332

Redenbaugh K, Fugii JA, Slade D (1988) Encapsulated plant embryos. In: Mizrah A (ed) Biotechnology in agriculture. Alan R Liss, New York, pp 225–248

Reed BM (1989) The effect of cold hardening and cooling rate on the survival of apical meristems of *Vaccinium* species frozen in liquid nitrogen. Cryo Lett 10:315–322

Reed BM (1992) Cryopreservation of *Ribes* apical meristems. Cryobiology 29:740

Reed BM, Lagerstedt HB (1987) Freeze preservation of apical meristems of *Rubus* in liquid nitrogen. HortScience 22:302–303

Seibert M, Wetherbee PJ (1977) Increased survival and differentiation of frozen plant organ cultures through cold treatment. Plant Physiol 59:1043–1046

Shimonishi K, Ishikawa M, Suzuki S, Oosawa K (1991) Cryopreservation of melon somatic embryos by a dessication method. Jpn J Breed 41:347–351

Taniguchi K, Tanaka R, Ashitani N, Miyagawa H (1988) Freeze preservation of tissue cultured shoot primordia of the annual *Haplopappus gracilis* (2n = 4). Jpn J Genet 63:267–272

Towill LE (1981) Survival at lowtemperature of shoot tips from cultivars of *Solanum tuberosum* group *tuberosum*. Cryo Lett 2:373–382

Towill LE (1991) Cryopreservation. In: Dodds JH (ed) In vitro methods for conservation of plant genetic resources. Chapman & Hall, London, pp 41–70

Uemura M, Sakai A (1980) Survival of carnation (*Dianthus caryophyllus* L.) shoot apices frozen to the temperature of liquid nitrogen. Plant Cell Physiol 21:85

Withers LA (1979) Freeze preservation of somatic embryos and clonal plantlets of carrot (*Daucus carota* L.) Plant Physiol 63:460–467

Withers LAW, Benson EE, Martin M (1988) Cooling rate/culture medium interactions in the survival and structural stability of cryopreserved shoot tips of *Brassica napus*. Cryo Lett 9:114–119

Yamada T, Sakai A, Matsumura T, Higuchi S (1991) Cryopreservation of apical meristems of white clover (*Trifolium repens* L.). Plant Sci 73:111–116

5 Storage of Free Pollen, Pollen Embryos and the Zygotic Embryos of Seed by Cryopreservation and Freeze Drying

B.W.W. Grout and A.V. Roberts

The value of stored pollen as an efficient way of conserving plant genes in vitro has been widely recognised, as part of a broader strategy for the conservation of plant genetic resources (Towill 1985; Withers and Williams 1986; Alexander and Ganeshan 1988; Withers et al. 1990; Withers 1991). Stored pollen, recovered with acceptable pollination competence, is also of great value in plant breeding and production, as an appropriately compiled inventory of such material removes many difficulties associated with differences in flowering time, season, location and availablility of partners that impair specific plant crosses. The increasing successes of in vitro technology for pollen culture, leading to tissue and plant regeneration, has advantages for industrial and agricultural biotechnology and, consequently, generates a further need for effective pollen storage (e.g. Alexander and Ganeshan 1988; Chen 1990; Tupy et al. 1991). Practical storage procedures, effective for months to several years in different species, have been developed for a wide range of pollens by simple reduction of environmental temperature and humidity, either separately or in combination (King 1965; Towill 1985; Barbosa et al. 1991; Yates et al. 1991; Niimi and Shiokawa 1992; Osborne et al. 1992). These storage periods can be significantly extended in those instances where vacuum- or freeze drying can be successfully applied to the mature pollen (King 1965; Akihama et al. 1978; Towill 1985), and techniques that provide storage for up to 4 years have been developed for a wide range of fruit and vegetable crops, ornamentals, forest and agricultural species.

Cryopreservation provides an additional dimension for pollen storage, offering the possibility of significantly extended storage periods with the maximum of genetic stability (see Grout, Chap. 1, this Vol.). This is particularly desirable when dealing with haploid material that has only a single representative of each chromosome. Further, the increasing understanding of the cryobiology underlying successful cryopreservation makes this a likely route for storage success with pollens that are problematic and do not freeze dry satisfactorily. As with all other tissue types there are inherent dangers in the eventual adoption of what may prove to be sub-optimal procedures for pollen conservation (Crisp and Grout 1984) and, during characterisation prior to storage, germination of pollen grains per se should not be viewed as an acceptable measure of success of the protocol. The production of seeds in acceptable numbers, with the expected characteristics, must be taken as the necessary measure of success.

Pollen embryos are organised structures, typically derived from cultured anthers, and with the gametic chromosome complement. They have the potential to develop into tissues and plantlets, and so are of considerable value in plant breeding and production (e.g. Bajaj 1984; Andersen et al. 1987; Chen 1990; Roulund et al. 1991). They are transient structures in vitro, following a developmental pathway in the same manner as somatic embryos (see Grout, Chap. 4, this Vol.) and their value and contribution to plant biotechnology can be greatly enhanced by suspending this development, whilst retaining viability, at an appropriate stage to provide long-term storage with the maximum of genetic stability. Cryopreservation, therefore, becomes the storage procedure of choice. Pollen embryos are also viewed as targets for DNA transfer using techniques such as microprojectile particle bombardment (Loe and Reynolds 1993), providing another application where cryopreservation is desirable.

Table 1. Examples of successfully cryopreserved pollen, pollen embryos and zygotic embryos

	Germination	Seed set	Reference
1] *Glycine max*	+	+	Collins et al. 1973
2] *Secale cereale*	+	+	" " "
3] *Trifolium praetense*	+	+	Engelke and Smith 1974
4] *Zea mays*	+		Barnabas and Rajki 1976
5] *Solanum tuberosum*	+		Weatherhead et al. 1978
6] *Brassica oleracea*		+	Crisp and Grout 1984
7] *Narcissus* sp.			Bowers 1990
8] *Helianthus annus*	+		Andreica and Sparchez 1990
9] *Phytolacca dodecandra*	+	+	Demeke and Hughes 1991
10] *Clianthus formosus*	+		Hughes et al. 1991
11] *Citrus limon*	+		Ganeshan and Alexander 1991
12] *Allium victorialis*	+		Kanazawa et al. 1992
13] *Arachis villosa*			Bajaj 1983
14] *Brassica napus*			" "
15] *Triticum aestivum*			Bajaj 1984
16] *Oryza sativa*			" "
17] *Triticum aestivum*			Bajaj 1984
18] *Oryza sativa*			" "
19] *Hevea brasiliensis*			Normah et al. 1986
20] *Artocarpus heterophyllus*			Chin et al. 1989
21] *Nephelium lappaceum*			Chin et al. 1989
22] *Manihot esculenta*			Marin et al. 1990
23] *Camellia sinensis*			Chaudhury et al. 1991
24] *Juglans regia*			de Boucaud et al. 1991

The genetic resources of many valuable crop plants, especially fruit and timber crops of tropical regions, are contained in recalcitrant seed which cannot be stored by desiccation or temperature reduction to subzero levels (Chin and Roberts 1980; Chin et al. 1984; Grout 1986; Withers 1987; Chin 1991). Typically, lethal injury to such seeds results from reduction of water content below 15–30% of fresh weight, depending upon species, and irreversible damage results from exposure to freezing temperatures. Consequently, extended storage of such materials is problematic, but a number of studies have indicated that the zygotic embryo, when excised from the entire seed, may show more orthodox behaviour and withstand moisture reduction to 10% of fresh weight or lower (Grout et al. 1983; Chin 1991 and see Table 5.1). This material, with appropriate treatment, is amenable to cryopreservation in the same way as embryos from seeds recognised as orthodox (Roberts et al. 1984) and requires only a successful in vitro embryo culture technique to provide a complete, practical system for storage.

There is also interest in the cryopreservation of the embryos of orthodox seeds, as an alternative to the storage of intact material, in situations where the embryo may be a source of embryogenic callus (Kendall et al. 1993) or secondary embryos (Chen 1990) and where immature embryos are involved following embryo rescue.

Objectives

The objectives of this chapter are to provide information

- on the practical aspects of storage of mature pollen by freeze-drying,
- on the practical aspects of cryopreservation of pollen, pollen embryos and zygotic embryos excised from seed,
- for the satisfactory assessment of the recovery of material from storage.

5.1 Freeze Drying of Pollen

The following protocol has proved to be successful for a relatively wide range of species (Withers 1987), and the reader is referred to King (1965), Akihama et al. (1978) and Towill (1985) for further details and wider applications of the technology.

Equipment and materials

- Freshly harvested pollen with a known germination capability and, if possible, a proven ability to set seed. This information is part of the characterisation process.

- A suitable system for desiccating pollen to a known moisture content, using either timed dehydration in a conventional dessicator or timed exposures to equilibrated atmospheres over solutions of known relative humidity.
- A suitable freeze-drying system including vials that will withstand exposure to liquid nitrogen.
- A flask of liquid nitrogen for plunging vials to achieve freezing prior to freeze drying a storage environment at −20 or −196 °C.
- A hygrostat at 90% RH to rehydrate pollen after recovery from the storage environment.
- A survival assay (Heslop-Harrison and Heslop-Harrison 1977) requiring a freshly made stock solution of fluoroscein diacetate (0.15% w/v in absolute acetone), stored at 4 °C. When required, a working solution is made by adding three drops of the stock solution to 10 ml of standard growth medium.
- A germination assay. A typical medium is 10–40% w/v sucrose with 50–200 ppm boron in distilled water (Towill 1985). This may be used in a liquid form or, more conveniently, as a gel with 0.9% w/v agar.

Procedures

It should be noted that incubation times given in the following text are guidelines only, and that empirical investigation will be required to determine the optimum times in any given instance.

1. The previous notes on the physical aspects of liquid nitrogen storage, the handling of liquid nitrogen, the sampling of populations for storage and the replication of stored cultures apply.
2. Reduce moisture content of the pollen to the required level.
3. Transfer small volumes of pollen to the vials and prefreeze by direct immersion in liquid nitrogen. This step can be considered as optional as it often confers no obvious benefits to the success of the procedure (King 1965; Towill 1985).
4. Move vials to the freeze dryer and dry for 30 min under vacuum.
5. Seal when process is complete and store in an appropriate low-temperature environment.
6. Thaw when required and rehydrate the pollen by transferring the opened vials to the hygrostat for an appropriate period.

7. Immediate survival is determined by placing a small sample of thawed pollen grains on a slide and adding two to three drops of the diluted fluoroscein diacetate solution. After a few minutes, the sample is carefully covered and examined under the microscope using UV light (excitation filter 355 nm, barrier filter 420 nm). Surviving grains will show bright, clear fluorescence and should be scored for five randomly selected fields of view for each sample. Use the lowest suitable magnification. A quantitative measure of survival, as a percentage, is given by:

   Number of fluorescent grains/total number of grains viewed $\times 100$.

8. Germination is determined by plating samples of recovered grains onto duplicate plates of the germination medium and incubating at 20 °C for 6 h before scoring for germinated grains.

5.2 Pollen Cryopreservation

The reader is specifically referred to entries 1–12 in Table 1.

Successful cryopreservation of pollen has typically used direct immersion of a vial containing the pollen sample into liquid nitrogen as a cooling technique. A great diversity of vials has been used and rarely have the achieved cooling rates been reported in the literatutre. It can be assumed that the rate will be in excess of 100 °C per min for all the commercial vials approved for cryogenic use.

Reduction in pollen water content prior to cooling is critical for survival in many instances (Towill 1985), and this is best achieved by pretreatment in a dessicator for a calculated period, or by equilibration to constant weight in the atmosphere generated above solutions of known relative humidity.

Equipment and materials

- Freshly harvested pollen with a known germination capability and, if possible, a proven ability to set seed. This information is part of the characterisation process.
- A suitable system for desiccating pollen to a known moisture content, using either timed dehydration in a conventional dessicator or timed exposures to equilibrated atmospheres over solutions of known relative humidity.
- Polypropylene vials suitable for cryogenic storage.
- Liquid nitrogen bath for plunging vials to achieve the cryogen temperature, and a liquid nitrogen vessel suitable for the long-term storage of biological materials.

- A thawing bath of sterile distilled water at 40 °C.
- A survival assay (Heslop-Harrison and Heslop-Harrison 1977) requiring a freshly-made stock solution of fluoroscein diacetate (0.15% w/v in absolute acetone), stored at 4 °C. When required a working solution is made by adding three drops of the stock solution to 10 ml of standard growth medium.
- A germination assay. A typical medium is 10–40% w/v sucrose with 50–200 ppm boron in distilled water (Towill 1985). This may be used in a liquid form or, more conveniently, as a gel with 0.9% w/v agar.

Procedures

1. The previous notes on the physical aspects of liquid nitrogen storage, the handling of liquid nitrogen, the sampling of populations for storage and the replication of stored cultures apply.
2. Pollen samples for cryogenic storage are desiccated to an appropriate level, empirically determined by previous experimentation. Levels above 20% are likely to be unsuccessful. No further cryopreservation treatment is required.
3. The prepared pollen is transferred to cryogenic vials (max. volume of pollen sample is 0.2 ml in a 2.0-ml volume tube).
4. The vials are frozen by direct immersion in liquid nitrogen and are transferred into storage after they have ceased to generate excessive bubbling of the nitrogen.
5. During storage, the level of liquid nitrogen in the storage vessel must be regularly monitored and replenished as appropriate, and good practice observed with regard to the maintenance and operation of the storage system (see Grout, Chap. 3, this Vol.)
6. The pollen is thawed by immersion of the vials in the thawing bath for 1.5 min.
7. Immediate survival is determined by placing a small sample of thawed pollen grains on a slide and adding two to three drops of the diluted fluoroscein diacetate solution. After a few minutes, the sample is carefully covered and examined under the microscope using UV light (excitation filter 355 nm, barrier filter 420 nm). Surviving grains will show bright, clear fluorescence and should be scored for five randomly selected fields of view for each sample. Use the lowest suitable magnification. A quantitative measure of survival, as a percentage, is given by:

   Number of fluorescent grains/total number of grains viewed × 100.

8. Germination is determined by plating samples of recovered grains onto duplicate plates of the germination medium and incubating at 20 °C for 6 h before scoring for germinated grains.

5.3 Cryopreservation of Pollen Embryos

The reader is specifically referred to entries 13–16 in Table 1.

If further investigation is required, as the method outlined below is ineffective, then anther-derived pollen embryos may be dealt with in a similar way to somatic embryos frozen in suspension, in the first instance.

Equipment and materials

- Media and cryoprotectants should be prepared taking note of the warnings and guidelines previously presented (see Grout, Chap. 3, this Vol.). The notes on the physical aspects of liquid nitrogen storage, the handling of liquid nitrogen, the sampling of populations for storage and the replication of stored cultures also apply.
- The cryoprotectant solution is 7% w/v DMSO, 5% w/v glycerol and 5% w/v sucrose in the appropriate growth medium.
- Polypropylene vials designed for liquid nitrogen storage. These should be labelled as described in Chapter 3, this Volume.
- Liquid nitrogen bath for plunging vials to achieve the cryogen temperature, and a liquid nitrogen vessel suitable for the long-term storage of biological materials.
- A thawing bath of sterile distilled water at 40 °C
- An appropriate recovery medium for the pollen embryos, solidified with 0.8% w/v agar.

Procedures

1. The previous notes on the physical aspects of liquid nitrogen storage, the handling of liquid nitrogen, the sampling of populations for storage and the replication of stored cultures apply.
2. Precultured, androgenic anthers at the required stage of development are cut into small sections, or pollen embryos are taken into suspension by gentle agitation of dissected anthers, both in standard growth medium. These are transferred onto ice.
3. An equal volume of precooled cryoprotectant solution is added to the suspensions, with gentle mixing, over 20 min and a total incubation in protectant of 1 h is given.

4. The prepared anther segment/embryo suspension is transferred to cryogenic vials.

5. The vials are frozen by direct immersion in liquid nitrogen and are transferred into storage after they have ceased to generate excessive bubbling of the nitrogen.

6. During storage the level of liquid nitrogen in the storage vessel must be regularly monitored and replenished as appropriate, and good practice observed with regard to the maintenance and operation of the storage system (see Grout, Chap. 3, this Vol.).

7. The vials are thawed by immersion in the thawing bath with gentle agitation until ice is no longer visible.

8. The thawed material is plated onto the surface of the regrowth medium, and surplus liquid carefully drawn off after 4h if required.

5.4 Cryopreservation of Excised, Zygotic Embryos

The reader is specifically referred to entries 17–24 in Table 1.

There are two basic procedures for the successful cryopreservation of excised, zygotic embryos. These are essentially with (A), and without (B), the addition of chemical cryoprotectant, and procedure (A) may be a more suitable choice for immature embryos.

Equipment and materials

– With the addition of chemical cryoprotectant (A)

- Media and cryoprotectants should be prepared taking note of the warnings and guidelines previously presented (see Grout, Chap. 3, this Vol.). The notes on the physical aspects of liquid nitrogen storage, the handling of liquid nitrogen, the sampling of populations for storage and the replication of stored cultures also apply.

- Cryoprotectant solutions are based on standard growth medium with one of two groups of protectants:

  – 10%w/v DMSO with sucrose in the range 4–10%w/v
  – 5M 1.2 propanediol with 20% w/v sucrose.

- Polypropylene vials (1–2.5ml volume) designed for liquid nitrogen storage. These should be labelled as described in Chapter 3, this Volume.

- Liquid nitrogen bath for plunging vials to achieve the cryogen temperature, and a liquid nitrogen vessel suitable for the long-term storage of biological materials.

- A thawing bath of sterile distilled water at 40 °C.
- An appropriate recovery medium for the excised embryos. The possible benefits of activated charcoal as a medium additive in attempts to minimise callusing during post-thaw growth should be considered (Withers 1987).

– Without the addition of chemical cryoprotectant (B)

- A suitable system for drying the embryos in the airflow of the laminar flow work station whilst monitoring weight loss.
- Polypropylene vials (1–2.5 ml volume) designed for liquid nitrogen storage. These should be labelled as described in Chapter 3, this volume.
- Liquid nitrogen bath for plunging vials to achieve the cryogen temperature, and a liquid nitrogen vessel suitable for the long-term storage of biological materials.
- A thawing bath of sterile distilled water at 40 °C.
- An appropriate recovery medium for the excised embryos. The possible benefits of activated charcoal as a medium additive in attempts to minimise callusing during post-thaw growth should be considered (Withers 1987).

Procedure A

1. The previous notes on the physical aspects of liquid nitrogen storage, the handling of liquid nitrogen, the sampling of populations for storage and the replication of stored cultures apply.
2. The isolated embryos are treated for an optimum time (typically 40 min-1 h) in cryoprotectant at 0 °C.
3. Five to ten embryos in a maximum of 0.5 ml of medium are placed in a cryogenic vial, which is frozen by direct immersion in liquid nitrogen and transferred into storage after it has ceased to generate excessive bubbling in the nitrogen.
4. During storage the level of liquid nitrogen in the storage vessel must be regularly monitored and replenished as appropriate, and good practice observed with regard to the maintenance and operation of the storage system (see Grout, Chap. 3, this Vol.).
5. The vials are thawed by immersion in the thawing bath with gentle agitation until ice is no longer visible.
6. The thawed embryos are transferred to the appropriate medium for regrowth, and monitored for extension and regeneration of the origi-

nal axis, callusing, secondary embryogenesis and adventitious plant regeneration.

Procedure B

1. The excised embryos are dried, by exposure to moving air in the laminar flow cabinet, to a water content below 15% of their fresh weight.
2. Five to ten dried embryos are placed in a cryogenic vial, which is frozen by direct immersion in liquid nitrogen and transferred into storage after it has ceased to generate excessive bubbling in the nitrogen.
3. During storage, the level of liquid nitrogen in the storage vessel must be regularly monitored and replenished as appropriate, and good practice observed with regard to the maintenance and operation of the storage system (see Grout, Chap. 3, this Vol.).
4. The vials are thawed by immersion in the thawing bath for 1.5 min.
5. The thawed embryos are transferred to the appropriate medium for regrowth, and monitored for extension and regeneration of the original axis, callusing, secondary embryogenesis and adventitious plant regeneration.

References

Akihama T, Omura M, Kozai I (1978) Further investigation of freeze-drying for deciduous tree pollen. In: Akihama T, Nakjima K (eds) Long-term preservation of favourable germplasm. Fruit Tree Research Station MAF, Japan, pp 1–7

Alexander MP, Ganeshan S (1988) The role of pollen cryobanks in genetic conservation. Genome 30:Suppl 1. 472

Andersen SB, Due IK, Olesen A (1987) The response of anther culture in genetically wide material of winter wheat (*Triticum aestivum* L.) Plant Breed 99:181–186

Andreica A, Sparchez C (1990) In vitro germination of sunflower pollen preserved at low temperatures. Stidii si Cercetari de Biologie. Ser Biol Veg 42:51–54

Bajaj YPS (1983) Regeneration of plants from pollen embryos of *Arachis, Brassica* and *Triticum* cryopreserved for one year. Curr Sci 52:484–486

Bajaj YPS (1984) The regeneration of plants from frozen pollen embryos and zygotic embryos of wheat and rice. Theor Appl Genet 67:525–528

Barbosa W, Campo-Dall'Orto FA, Ojima M, Martins FP, Boaventura YMS (1991) Pollen storage and germination, pollination and fruit set in sub-tropical peaches and nectarines. Bragantia 50:17–28

Barnabas B, Rajki E (1976) Storage of maize (*Zea mays* L.) pollen at −196 °C in liquid nitrogen. Euphytica 25:747–752

Bowers SA (1990) Long-term storage of *Narcissus* anthers and pollen in liquid nitrogen. Euphytica 48:275–278

Chaudhury R, Radhamani J, Chandel KPS (1991) Preliminary observations on the cryopreservation of desiccated embryonic axes of tea [*Camellia sinensis* (L.) O. Kuntze] seeds for genetic conservation. Cryo Lett 12:31–36

Chen Z (1990) Haploid induction in perennial crops. In: Chen Z, Evans DA, Sharp WR, Ammirato PV, Sondahl MR (eds) Handbook of plant cell culture, vol 6. Perennial crops. McGraw-Hill, New York, pp 62–75

Chin HF (1991) Conservation of recalcitrant seeds. In: Zakri AH, Normah MN, Senawi MT, Abdul Karim AG (eds) Conservation of plant genetic resources through in vitro methods. FRIM/MNCPGR, Kuala Lumpur, pp 19–27

Chin HF, Roberts HE (1980) Recalcitrant crop seeds. Tropical Press, Kuala Lumpur

Chin HF, Hor YL, Mohd Lassim MB (1984) Identification of recalcitrant seeds. Seed Sci Technol 12:429–436

Chin HF, Krishnapillay B, Hor YL (1989) A note on the cryopreservation of embryos from young coconuts (*Cocos nucifera* var. Mawa). Pertanika 12:183–186

Collins FC, Lertmongkol V, Jones JP (1973) Pollen storage of certain agronomic species in liquid air. Crop Sci 13:493–498

Crisp PC, Grout BWW (1984) Storage of broccoli pollen in liquid nitrogen. Euphytica 33:819–823

de Boucaud M, Brison M, Ledoux C, Germain E, Lutz A (1991) Cryopreservation of embryonic axes of recalcitrant seed of *Juglans regia* L. cv Franquette. Cryo Lett 12:163–166

Demeke T, Hughes HG (1991) Germination and storage of pollen of *Phytolacca dodecandra* L. Ann Bot 68:13–16

Engelke MC, Smith RR (1974) Effect of storage on the germination and viability of red clover pollen grains. Agron Abstr (Am Soc Agron Madison) 1974:52

Ganeshan S, Alexander MP (1991) Cryogenic preservation of lemon (*Citrus limon* Burm.) pollen. Gartenbauwissenschaft 56:228–230

Grout BWW (1986) Embryo culture and cryopreservation for the conservation of genetic resources of species with recalcitrant seed. In: Withers LA, Alderson PG (eds) Plant tissue culture and its agricultural applications. Butterworths, London, pp 303–309

Grout BWW, Shelton K, Pritchard HW (1983) Orthodox behaviour of oil palm seed and cryopreservation of the excised embryo for genetic conservation. Ann Bot 52: 381–384

Heslop-Harrison J, Heslop-Harrison Y (1977) Evaluation of pollen viability by enzymatically-induced fluorescence: intracellular hydrolysis of fluoroscein diacetate. Stain Technol 45:115–121

Hughes HG, Lee CW, Towill LE (1991) Low temperature preservation of *Clianthus formosus* pollen. HortScience 26:1411–1412

Kanazawa T, Kobayashi S, Yakuwa T (1992) The flowering process, germination and storage of pollen in *Allium victorialis* ss. *platyphyllum*. J Jpn Soc Hortic. Sci 60:947–953

Kendall EJ, Kartha KK, Qureshi JA, Chermak P (1993) Cryopreservation of immature spring wheat zygotic embryos using an abscisic acid pretreatment. Plant Cell Rep 12:89–94

King JR (1965) The storage of pollen – particularly by the freeze-drying method. Bull Tortey Bot Club 92:270–287

Loeb TA, Reynolds TC (1993) Transient expression from microprojectile-mediated DNA transfer in pollen embryos of bread wheat. Plant Physiol 102:Suppl 165

Marin ML, Mafla G, Roca WM, Withers LA (1990) Cryopreservation of *Cassava* zygotic embryos and whole seeds in liquid nitrogen. Cryo Lett 11:257–264

Niimy Y, Shiokawa Yu (1992) A study on the storage of *Lilium* pollen. J Jpn Soc Hortic Sci 61:393–403

Normah MN, Chin HF, Hor LY (1986) Desiccation and cryopreservation of embryonic axes of *Hevea brasiliensis* Muell-Arg. Pertanika 9:299–303

Osborne R, Robbertse PJ, Claassen MI (1992) The longevity of cycad pollen in storage. S Afr J Bot 58:250–254

Roberts EH, King MW, Ellis RH (1984) Recalcitrant seeds – their recognition and storage. In: Holden JWH, Williams JT (eds) Crop genetic resources: conservation and evaluation. Allen & Unwin, London, pp 38–52

Roulund N, Andersen SB, Farestveit B (1991) Optimal concentration of sucrose for head cabbage [*Brassica oleracea* L. convar. capitata (L.) Alef] anther culture. Euphytica 52:125–129

Towill LE (1985) Low temperature and freeze-/vacuum-drying preservation of pollen. In: Kartha KK (ed) Cryopreservation of plant cells and organs. CRC Press, Boca Raton, pp 171–198

Tupy J, Rihova L, Zarsky V (1991) Production of fertile tobacco pollen from microspores in suspension culture and its storage for in situ pollination. Sex Plant Reprod 4:284–287

Weatherhead MA, Grout BWW, Henshaw GG (1978) Advantages of the storage of potato pollen in liquid nitrogen. Potato Res 21:97–100

Withers LA (1987) The low temperature preservation of plant cell, tissue and organ cultures and seed for genetic conservation and improved agricultural practice. In: Grout BWW, Morris GJ (eds) The effects of low temperatures on biological systems. Edward Arnold, London, pp 389–409

Withers LA (1991) Tissue culture in the conservation of plant genetic resources. In: Zakri AH, Normah MN, Senawi MT, Abdul Karim AG (eds) Conservation of plant genetic resources through in vitro methods. FRIM/MNCPGR, Kuala Lumpur, pp 1–18

Withers LA, Williams JT (1986) In vitro conservation – international board for plant genetic resources research highlights 1985–1985. IBPGR, Rome

Withers LA, Wheelans SK, Williams JT (1990) In vitro conservation of crop germplasm and the IBPGR databases. Euphytica 45:9–22

Yates E, Sparks D, Connor K, Towill L (1991) Reducing pollen moisture simplifies long-term storage of pecan pollen. J Am Soc Hortic Sci 116:430–434

6 Conservation of Algae

J.G. Day and M.R. McLellan

Algae, which in the context of this chapter include both eukaryotic microalgae and prokaryotic cyanobacteria, are virtually ubiquitous in euphotic aquatic environments and are commonly found in terrestrial environments. Some strains are capable of living in extremely ecologically stressful environments, e.g. endolithic cyanobacteria "desert varnish", which live beneath the surface of rocks (Friedman 1980). Others can form symbiotic associations with fungi (Stewart et al. 1983), liverworts (Rodgers and Stewart 1977), higher plants (Rai et al. 1984), protozoa (Wilcox and Wedemayer 1985) and aquatic invertebrates (Reisser 1985). As with other organisms, the requirement for authentic specimens for use in research and other applications has led to the development of maintenance procedures for the retention of live specimens. Simultaneously, there has been a development of algal culture collections, which vary in size from small specialist collections, for example in aquaculture and ecotoxicity testing centres, to larger specialist collections with several hundred strains, e.g. The Pasteur Culture Collection of Cyanobacterial Strains, up to the largest service collections (Table 1). It has been estimated that there are approximately 50,000 species of microalgae of which approximately 2.5% are currently lodged in collections (Hawksworth and Round 1991). This figure probably represents one of the highest proportion of any group of microorganisms to be retained in culture and lodged in official collections. The concerns regarding the reduction of biodiversity, the development of new isolation techniques (O'Donnell 1992) and the growth of molecular systematics (Olsen 1990) will undoubtedly lead to the isolation, identification and preservation of a larger proportion of the algal flora.

Conservation technique development for algae has been less intensively researched than for other major groups of microorganisms. Techniques can be split into two groups; serial subculture, and long-term preservation techniques.

Most algae are photoautotrophic in nature; however, some are capable of utilizing organic substrates, and other achlorophyllous species are obligate heterotrophs. This physiological diversity and the variety of environments from which they have been isolated have given rise to a multitude of different media and culture conditions and techniques. This chapter gives in some

Table 1. List of major algal culture collections. (After Day and Turner 1992)

Acronym	Name	Country	No. of Strains
ASIB	Algensammlung am Institut für Botanik	Austria	1570
CALU	Collection of Algal Cultures Leningrad Univ	USSR	600
CCALA	Culture Collection of Autotrophic Organisms	Czech Republic	498
CCAP	Culture Collection of Algae and Protozoa	UK	1576
CCMP	Provasoli-Guillard Center for Culture of Marine Phytoplankton	USA	1000
IAM	Institute of Applied Microbiology	Japan	500
NIES	Microbial Culture Collection	Japan	500
SAG	Sammlung von Algenkulturen	Germany	1400
UTEX	Culture Collection of Algae at the Univ. Texas at Austin	USA	2089

Major collection = collection with >450 cultures lodged.

detail the protocols and media employed to maintain the majority of algal strains currently retained in culture.

The need for genetic conservation, and the selective nature, expense and labour intensity of routine serial subculture have led to the development of a number of methods of long-term preservation of algae of varying suitability. Lyophilization and air drying have largely been found to be unsuitable, with low levels of viability (0–1%) on storage for 1 year (Day 1992). However, improved drying techniques have been developed which could potentially be used for the medium to long-term storage of some algae (Malik 1992). Cryopreservation is undoubtedly the most effective method of long-term preservation, as maintenance costs for frozen cultures are very low and, assuming the storage vessel temperature is maintained at below −135 °C, viability is independent of storage duration (Morris 1981). There is, as yet, no universal methodology for the cryopreservation of algal species which assures high levels of post-thaw viability. The most commonly used successful protocols are described in Section 6.2.

6.1 Serial Subculture

Objectives

- To retain a healthy, physiologically, morphologically and genetically representative specimen.
- To maintain a metabolically active culture which can be used at short notice.

The main limitations of this technique are the selective and artificial nature of the media and incubation regimes with respect to native ecological conditions. Laboratory conditions can, in extreme cases, lead to the loss of important morphological features and physiological traits. Further potential problems include the possibility of contamination of axenic cultures and the possibility of mislabelling or other handling mistakes.

Equipment and materials

The following equipment and facilities are required:

- Appropriate sterilizable, culture vessels
- tubes
- flasks
- aspirators
- facilities for aseptic transfers
- laminar-flow cabinet
- clean room
- Incubators or constant temperature rooms capable of temperature control in the range 5–35 °C and light /dark cycles in the range 8:16h–20:4h

Culture media

The media are listed under acronyms commonly used in the literature. The following points should be noted before preparing media.

- Stock solutions must not be combined before dilution, otherwise precipitation may occur. Always add stock solutions to a large proportion of the final volume. Stock solutions can be stored for prolonged periods at 5 °C, but should be discarded if obviously contaminated or if precipitate has formed.
- Sterilization is by autoclaving at 121 °C (15 lb in^{-2}; ca. 1 bar or 101 kPa) for 15 min or longer if large volumes are required.
- Ultramarine Synthetica sea salts are available from Waterlife Research Industries Limited, 476 Bath Road, Longford, West Drayton, Middlesex UB7 0ED, England. Alternative proprietary brands of artificial sea salts may be substituted if necessary.
- Soil extract (SE): mix 1 part air-dried, sieved soil with two parts distilled water. Adjust to pH 8.0 with NaOH or HCl and autoclave for 1 h at 15 lb in^{-2} pressure. Decant or filter the supernatant.
- Solidified media are produced on addition of 10–15 g l^{-1} of agar (Oxoid No. 1) to the media listed below. Test tubes, universals or other appropriate vessels are placed at an angle of approximately 30° during agar gelation, to maximise the surface area available for growth.

Freshwater media:

Table 2. BG 11 [blue-green algal (cyanobacterial) medium]. (After Thompson et al. 1988)

Stock solutions	$g\,l^{-1}$
1. $NaNO_3$	15.0[a]
2. $K_2HPO_4.3H_2O$	4.0
3. $MgSO_4.7H_2O$	7.5
4. $CaCl_2.2H_2O$	3.6
5. Citric acid	0.6
6. Ferric ammonium citrate	0.6
7. EDTA (disodium salt)	0.1
8. Na_2CO_3	2.0
9. Trace metal mixture:	
H_3BO_3	2.86
$MnCl_2.4H_2O$	1.81
$ZnSO_4.7H_2O$	0.222
$Na_2MoO_4.2H_2O$	0.39
$CuSO_4.5H_2O$	0.079
$Co(NO_3)_2.6H_2O$	0.0494

[a] May be omitted for nitrogen-fixing Cyanobacteria.
Add 100 ml of solution 1, 10 ml of solutions 2–8 and 1 ml of solution 9 to distilled or deionised water to obtain a total volume of 1 l. pH is adjusted to 7.8 prior to sterilization.

Table 3. DM (diatom medium). (After Thompson et al. 1988)

Stock solutions	g per 200 ml
1. $Ca(NO_3)_2.4H_2O$	4.0
2. KH_2PO_4	2.48
3. $MgSO_4.7H_2O$	5.0
4. $NaHCO_3$	3.18
5. EDTA FeNa	0.45
EDTA Na_2	0.45
6. H_3BO_3	0.496
$MnCl_2.4H_2O$	0.278
$(NH_4)_6Mo_7O_{24}.4H_2O$	0.2
7. Cyanocobalamin (Vitamin B12)	0.008
Thiamine HC1 (Vitamin B1)	0.008
Biotin	0.008
8. $Na_2SiO_3.9H_2O$	11.4

Add 1 ml of solutions 1–8 to distilled or deionized water to obtain a total volume of 1 l. Add approximately 0.3 ml of 1 M HCl to adjust pH to 6.9 prior to sterilization.

Table 4. EGM (*Euglena gracilis* medium). (After Thompson et al. 1988)

	per litre
Sodium acetate (trihydrate)	1.0[a]
Lab-Lemco powder (Oxoid L29)	1.0
Tryptone	2.0
Yeast extract	2.0
$CaCl_2$ stock solution ($1.0\,g\,l^{-1}$)	10.0 ml

[a] For some achlorophyllous heterotrophs substitute glucose at $1\,g\,l^{-1}$.

Table 5. JM (Jaworski's medium). (After Thompson et al. 1988)

Stock solutions	g per 200 ml
1. $Ca(NO_3)_2.4H_2O$	4.0
2. KH_2PO_4	2.48
3. $MgSO_4.7H_2O$	10.0
4. $NaHCO_3$	3.18
5. EDTA FeNa	0.45
EDTA Na_2	0.45
6. H_3BO_3	0.496
$MnCl_2.4H_2O$	0.278
$(NH_4)_6Mo_7O_{24}.4H_2O$	0.2
7. Cyanocobalamin (vitamin B12)	0.008
Thiamine HCl (vitamin B1)	0.008
Biotin	0.008
8. $NaNO_3$	16.0
9. $Na_2HPO_4.12H_2O$	7.2

Add 1 ml of solutions 1–9 to distilled or deionised water to obtain a total volume of 1 l of medium.

Table 6. SES (soil extract medium with additional salts). (After Thompson et al. 1988)

Stock solutions	$g\,l^{-1}$
1. K_2HPO_4	1.0
2. $MgSO_4.7H_2O$	1.0
3. KNO_3	10
4. SE	

Add 20 ml of solutions 1–3 to 100 ml of SE and make up to 1 l with distilled or deionised water.

Marine media:

Table 7. ASW (artificial sea water). (After Thompson et al. 1988)

Stock solutions	per litre
1. $NaNO_3$	30.0 g
Na_2HPO_4	1.2 g
K_2HPO_4	1.0 g
2. Biotin	0.2 mg
Calcium pantothenate	20.0 mg
Cyanocobalamin (vitamin B12)	4.0 mg
Folic acid	0.4 mg
Inositol	1000.0 mg
Nicotinic acid	20.0 mg
Thiamine HCl (vitamin B1)	100.0 mg
Thymine	600.0 mg
3. Soil extract (SE)	

Dissolve 33.6 g of Ultramarine synthetic sea salts and 0.5 g of Tricine in 900 ml of distilled or deionised water. To the above add 25 ml soil extract and 3.75 ml solution 1. Make up to 1 l with distilled water. Adjust pH to 7.6–7.8 with 1 N NaOH or 1 N HCl, prior to sterilisation.

The vitamin solution (solution 2) is added after sterilisation through a 0.22 μm filter, 2.5 ml l^{-1}. Alternatively, vitamins may be sterilised in situ if vitamin concentration is not critical.

Variation: double strength ASW.

Some strains require sodium chloride to be added to the medium. Follow the recipe for ASW but add 35 g NaCl l^{-1} to the major salts.

Table 8. E27 (E27 medium). (After Thompson et al. 1988)

Stock solutions	per litre
1. KNO_3	25 g
2. $MgSO_4.7H_2O$	2.5 g
3. K_2HPO_4	2.5 g
4. Cyanocobalamin (vitamin B12)	0.1 mg
Thiamine HCl (vitamin B1)	50 mg
5. Soil Extract (SE)	

Liver extract (0.125 g), tryptone (0.125 g) and glucose (0.125 g) are dissolved in 500 ml of distilled or deionized water to which 1 ml of solutions 1–4 and 12.5 ml of solution 5 is added; the mixture is then autoclaved. On cooling either 500 ml of autoclaved filtered natural seawater or 17.5 g of Ultramarine synthetic sea salts in 500 ml of distilled water, presterilised by filtration or autoclaving, are added.

Table 9. GUI (Guillard's medium f/2). (After Thompson et al. 1988)

Stock solutions	per litre
1. $NaNO_3$	7.5 g
2. $NaH_2PO_4.2H_2O$	5.65 g
3. Na_2EDTA	4.36 g
$FeCl_3.6H_2O$	3.15 g
$CuSO_4.5H_2O$	0.01 g
$ZnSO_4.7H_2O$	0.022 g
$CoCl_2.6H_2O$	0.01 g
$MnCl_2.4H_2O$	0.18 g
$Na_2MoO_4.2H_2O$	0.006 g
4. Cyanocobalamin (vitamin B12)	0.5 mg
Thiamine HCl (vitamin B1)	100.0 mg
Biotin	0.5 mg
5. $Na_2SiO_3.9H_2O$	40.18 g

Add 10 ml of solution 1 and 1 ml of solutions 2–4 to filtered natural seawater to obtain a total volume of 1 litre. Adjust pH to 8.0 prior to sterilization, 1 ml of solution 5 is added to the medium prior to pH adjustment and sterilization, if the medium is to be used for diatoms.

- Mixtures of media are commonly used, generally in the proportion 1:1, unless otherwise stated.

Procedure

6.1.1 Sub-Culturing

Sub-culturing is performed using aseptic microbiological technique and involves transferring an inoculum (1–10% of the original culture) from a late log/stationary phase culture into fresh, pre-sterilized medium. Intervals between transfers are generally between 2 weeks and 6 months, depending on the growth rate of the organism and the incubation conditions. Stock cultures are usually maintained in test tubes capped with siliconised rubber or cottonwool bungs or glass universal tubes, in either liquid medium or on agar slopes.

6.1.2 Maintenance

The successful maintenance of algae is dependant on the selection of an appropriate medium and growth conditions including temperature and light intensity. These are discussed in more detail below.

Medium Selection. The choice of medium is dependent on the nutritional requirements and environmental origin of the alga, as well as its morphological and taxonomic characteristics. Many strains will grow satisfactorily on a variety of media and both in liquid and on solidified media. Others have specific requirements for vitamins or other nutrients. In general, the composition of defined and semi-defined media ensures that nutrient levels are well in excess of those in natural environments and that trace elements and other "growth factors" are not limiting. Richly organic media are used only for axenic strains, whereas contaminated, uni-algal cultures are retained on basic mineral medium, or medium with low levels of organics in the form of vitamins or humic acid. Table 10 gives an outline of media usage. For the culturing of organisms from extreme environments, specialist literature should be consulted (Starr and Zeikus 1987; Thompson et al. 1988; Nerad 1991; Watanabe and Satake 1991).

Temperature. Optimal growth temperatures vary, most being in the range 10–25 °C, with many algae having relatively broad growth optima. It is often desirable to grow some strains under sub-optimal conditions, in order to extend the sub-culture interval. Most algae can be incubated at 15 °C, although it is useful to maintain others at temperatures as low as 10 °C. Organisms isolated from extreme environments will normally grow satisfactorily at 15–20 °C, although it should be noted that this is imposing an additional selective influence on the alga.

Table 10. Suitability of media for different algal groups

Algal group	Axenic	Freshwater	Marine
Cyanophyta	+/–[a][b]	BG 11, JM	ASW:BG 11
Chlorophyta	+	EGM, EGM:JM, JM	ASW, E27, GUI
Chlorophyta	–	BG 11, DM, JM	ASW, GUI
Euglenophyta	+	EGM,EGM:JM	–
Euglenophyta	–	JM, JM:SES	ASW:SES (1:3)
Phaeophyta	–	DM	ASW, GUI
Chrysophyta	–	DM, JM	ASW, GUI
Bacillariophyceae	+	DM	E27, GUI+Si
Bacillariophyceae	–	DM	GUI+Si
Pyrrophyta	+	–	E27, GUI
Pyrrophyta	–	DM	ASW, GUI
Rhodophyta	+/–	DM, JM	ASW, GUI
Cryptophyta	–	JM	ASW, GUI
Achlorophyllous heterotrophs	+	EGM	E27

[a] + = axenic strains.
[b] – = unialgal strains.

Light. Artificial light provided by cool white or warm white fluorescent tubes is the standard method of providing illumination for photoautotrophic and mixotrophic algal cultures. Light should be provided on a light: dark cycle, 16:8 h and 12:12 h being the most commonly employed. Alternatively, natural light can be used, with cultures being placed in a north-facing window. Light intensity can be critical; bright light will cause severe physiological stress and eventual death in all algae. A light intensity of $50\,\mu E\,m^{-2}\,s^{-1}$ can be used for most groups of algae. Cyanobacteria require lower light levels, in the region of $25\,\mu E\,m^{-2}\,s^{-1}$ being optimal and many diatoms thrive at $200\,\mu E\,m^{-2}\,s^{-1}$.

Safety note

Although algae as a group are not widely considered to be pathogenic/harmful, due care should be taken on handling cultures. Some cyanobacteria and dinoflagellates contain potent hepato- and neuro-toxins (Collins 1978) and a number of achlorophyllous algae have been reported to be opportunistic pathogens (Fetter et al. 1971; Anonymous 1991).

Maintenance

The culture obtained should be identical to the parent culture and the original isolate. In reality, the best that can be achieved is to obtain a culture which is apparently morphologically and physiologically identical to the parent culture, since the original isolation procedure and the maintenance protocol are by their nature selective. Also the practice of using clonal cultures may select against a particularly important genetic, physiological or morphological trait. It cannot, therefore, be assumed that algae in culture are genetically identical to field samples.

Trouble shooting

Poor/no regeneration of the culture on transfer can usually be attributed to the inoculum, the medium or the incubation condition.

If possible, it is important to avoid using a senescent culture as inoculum. It may be possible to use a previous sub-culture or a duplicate sample which is more physiologically active than the culture initially chosen. If there is no option, a larger inoculum can be used, or alternatively, the nutrient regime and culture conditions can be altered (see below).

The most common problems associated with algal media are detailed below.

- *Incorrect pH*: most algae are tolerant of fairly large changes in pH; however, if the inoculum vigour is sub-optimal, poor/no growth can result. In most cases, freshwater eukaryotic algae prefer acidic environments (pH 5–7) whilst cyanobacteria prefer alkaline environments (pH 7–9).
- *High levels of precipitate*: these can result in nutrient limitation and osmotically stressful microenvironments.
- Incorrectly formulated *media*, or omission of a vital ingredient, e.g. Si for diatoms, vitamins for many marine algae.

- For *poor inocula*, improved recovery may be obtained by using a less defined medium in combination with the standard mineral medium, e.g. EGM:JM; ASW:SES, alternatively for non-axenic strains, supplementation of the medium with low concentrations of proteose peptone, vitamins or SE may assist recovery.

- The incubation *temperature* is unlikely to be the cause of growth failure, assuming the parent culture was originally grown under the same conditions. It is widely recognised that transferring cultures into an incubator set at a higher, physiologically suitable, temperature will stimulate the growth rate. This is often used to ensure a dense culture for rapid dispatch from collections, including CCAP, where slow-growing members of the Cyanophyta and Chlorophyta, normally incubated at 15°C, may be grown at 23°C.

- Incorrect *light levels* may be the cause of poor/no growth. Cyanobacterial cultures are particularly susceptible to photoinhibition, bleaching and death attributed to excessive light levels. It should be noted that in dense cultures, self-shading can have a significant effect on the intensity of light reaching an individual cell; sub-cultures therefore need to be initially incubated at relatively low light levels.

  Low light levels can have a limiting effect on final culture density, but assuming they were sufficiently high to support the growth of the parent culture, they are unlikely to prevent growth of the sub-culture.

- Localized *heating effects*: lights in close proximity to the culture may result in the culture temperature being above the temperature maximum for the strain. A common problem when illuminated incubators are not available occurs when cultures are placed in direct sunlight. Unless sufficient care is taken to control the intensity of light reaching the culture and the heating effects of the sun, the culture will invariably be killed.

- Use of *unsuitable lighting* regime: assuming that an appropriate light:dark regime is used, this is unlikely to cause problems. It is worth checking that the timing/switching mechanism is functioning effectively. Tungsten filament lights are not generally satisfactory, only fluorescent lights, GrowLux lights or natural light should be used.

Quality control

This is largely subjective, and is based on gross culture appearance and microscopic observation of cell morphology. If both appear normal, i.e. as for the parent culture or as in the scientific literature, this is assumed to be satisfactory.

For axenic cultures, purity checks involve microscopic examination to check for gross contamination plus aseptic transfer of small inocula from the parent culture into nutrient broth and spread onto plates of nutrient and tryptone soya agar. These are labelled appropriately and incubated for up to 10 days at 23 °C.

6.2 Cryopreservation

Objectives

- To maintain a viable, representative, genetically stable specimen.
- To have high levels of viability, in order to prevent selection of a freeze tolerant sub-population, and to minimise the time required post-thaw to regenerate a healthy, physiologically active, normal culture.
- To minimise the selective nature of in vitro conservation.
- To have no reduction in viability or alteration of the alga during prolonged storage.
- To minimise the maintenance cost of strains which are not regularly required.

The major limitations of this method of preservation are the requirement for specialist equipment (see Equipment and Materials, below), which may not be available in many laboratories, and the requirement for high levels of post-thaw viability. A universal protocol has not been developed, and currently cryopreservation at CCAP is only widely employed to preserve members of the Cyanophyta and Chlorophyta.

As detailed in Section 6.1.

Equipment and materials

- Cryoprotectants: dimethyl sulphoxide (DMSO), glycerol and methanol are the most commonly used.
- Sterile screw-top ampoules: 2-ml Costar Cryovials have been found to be most satisfactory, as their configuration incorporating an internal O-ring minimises the likelihood of liquid nitrogen contaminating the culture. The vials can be labelled using a suitable permanent pen, or colour-coded using a colour insert located in the tube cap.
- Refrigeration systems: a refrigerated immersion bath capable of maintaining a constant temperature in the range −25 to −40 °C, e.g. Fryka Kältetechnik KB 650 KR. A controlled-rate freezer, e.g. Planer Kryo 10.
- Liquid nitrogen.
- Safety equipment gloves, goggles or a face mask.

- Wide neck and standard format Dewars and an appropriate storage system which can be either mains-driven freezers, e.g. Gallenkamp Supercold 135 or more usually, liquid nitrogen, liquid phase (−196 °C) or gas phase (−135 °C) storage which in turn can be either an autofill system, e.g. Planer MV XLC 810 or a simple storage Dewar, e.g. Planer XLC 230. All the storage systems need to contain labelled inventories, suitable for retaining cryovials.
- A heated water bath (35–45 °C) suitable for thawing vials.

Procedure

6.2.1 Pre-Freezing Treatment

It has been demonstrated for some strains that environmental stress including prolonged incubation at 4 °C (Lesson et al. 1984), and nutrient limitations including; nitrate (Morris et al. 1980), bicarbonate (Ben-Amotz and Gilboa 1980) or silica (McLellan 1989) can increase levels of post-thaw survival of some algal strains.

At CCAP, late log-phase/early stationary phase cultures are frozen, as these tend to have the highest levels of post-thaw viability and are relatively dense, thus obviating the requirement for any concentration step.

6.2.2 Cryoprotective Addititives

A small number of algae can be preserved in the absence of any cryoprotective agent. In general, DMSO or glycerol at either 5 or 10% (v/v) are used, with methanol used for a number of *Chlorella* and *Euglena* strains. Cryoprotectants are made double strength, i.e. 10 or 20% (v/v) in appropriate medium and sterilized by autoclaving (DMSO, glycerol) or filtration (methanol). Medium containing cryoprotectant is mixed 1:1 with the algal culture and incubated for between 5 and 30 min at room temperature prior to freezing. This can be performed in the cryovial or in a flask prior to aseptically dispensing 1 ml of culture containing cryoprotectant into each vial.

6.2.3 Freezing Protocols

A variety of protocols have been developed for the preservation of algae (Morris 1981). These aim to minimise the damage induced by direct low temperature effects, including chilling and cold shock injury; freeze-induced exposure to hypertonic solutions; mechanical and physical stress due to ice

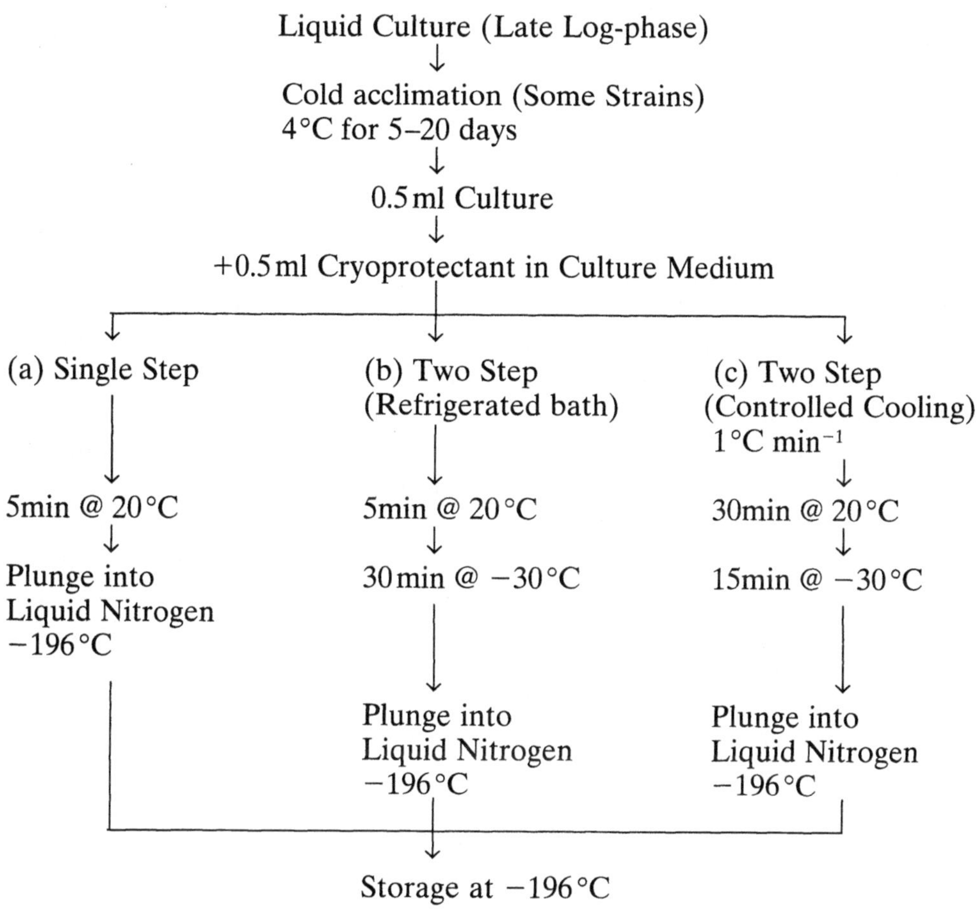

Fig. 1. Outline of cryopreservation protocols used at CCAP

formation, and the alteration of the properties of extra- and intracellular solutions. In reality, only a few protocols are regularly employed in culture collections and these are outlined in Fig 1. The two-step method (Fig. 1b) with DMSO (5% v/v) as a cryoprotectant is used to freeze the majority of strains currently preserved at CCAP.

6.2.4 Storage

Frozen samples, 1 ml in 2 ml cryovials, are transported to the storage refrigerator in a small Dewar of liquid nitrogen. Storage in any of the systems described under Equipment and materials is suitable, with a complete stock list being maintained at all times. It is a sensible precaution to ensure liquid nitrogen contact with the inventory rack at all times, irrespective of whether the system is maintaining the vials in the gas or liquid phase, as it is possible for the top of large refrigerator systems to reach temperatures above −130 °C, which may reduce long-term viability.

6.2.5 Thawing

Vials are transferred from the storage system into a wide-neck Dewar containing liquid nitrogen, and are then placed in a water bath at 40 °C and incubated until the ice has completely thawed. Effectively, this results in a thawing rate of 90 °C min^{-1}. The thawed culture is then used as an inoculum and aseptically transferred into fresh medium (9–50 ml) with no additional cryoprotectants and incubated as detailed in Section 6.1.2.

Safety note

Commonly employed cryoprotectants are often cytotoxic and potentially mutagenic at high concentrations. Care should be taken when dispensing these, and skin contact avoided. Liquid nitrogen should be handled with care and protective gloves and safety glasses worn at all transfers. Adequate ventilation, to prevent nitrogen gas build-up, must be ensured, and staff trained in the use of pressurised storage vessels and all relevant safety and first aid precautions.

Efficacy

A successful cryopreservation protocol should guarantee high levels of post-thaw viability and no reduction in viability during prolonged storage. Furthermore, thawed cells should recover rapidly and produce a culture identical to the parent culture.

The techniques described above can result in levels of viability of virtually 100% (Morris 1978) and viability levels are not significantly different on storage for up to 13 years (McLellan 1989).

When a cryopreserved cell is thawed, its physiological state may not be equivalent to that prior to freezing. A period of recovery may ensue post-thaw, where minor repair mechanisms are effected, and metabolic balance is achieved (Grout et al. 1978). Consequently, a longer lag phase may be expected in thawed algal cultures, compared to a similar inoculum density transferred from a non-frozen culture.

The possibility of sub-lethal chromosomal damage as a result of freeze/thaw cycles has been investigated for a number of yeast strains, and permanent and transient loss of plasmid activity has been observed (Pearson et al. 1990). Genome damage in cryopreserved algal strains, as evidenced by loss of functional characteristics, has not been reported, although further studies are required.

The following problems may occur:

Trouble shooting

- *Poor/no growth* post-thaw may be due to one or more of the factors listed below:

- Inappropriate protocol: check the relevant literature and "in house" information, or use the existing protocol as a starting point to develop a more satisfactory system.
- Poor inoculum quality: as detailed above, late-log/stationary phase cultures are widely used for preservation. Senescent or severely stressed cultures are unsatisfactory, with potentially high levels of cell lysis on freezing and low/no viability on thawing.
- If using a refrigerated bath system, check temperature calibration and that temperature levels are uniform throughout the bath.
- Possibility of partial thawing/refreezing during freezing/thawing procedures.
- Excessive period at intermediate holding temperature (−30 °C) for cultures frozen using a two-step protocol.
- Heat damage caused by excessive time spent in 40 °C water bath prior to transfer into fresh medium.
- Insufficient dilution of cryoprotectant may be a problem, particularly when DMSO or methanol are used as cryoprotectants.
- The points detailed under Trouble Shooting in Section 6.1 related to media and incubation conditions are equally relevant to post-thaw recovery.
- Improved recovery may be obtained by supplementing mineral media with low levels of organic material, e.g. yeast extract, proteose peptone or vitamins.

• *Contamination* of a formerly axenic culture may be due to a handling or transfer mistake. Alternatively, it may be due to leakage of liquid nitrogen into the vial. This is also potentially dangerous, as the nitrogen evaporates during thawing and could theoretically result in the vial exploding. For this reason, glass vials should be avoided and extreme caution exercised when thawing vials containing liquid nitrogen. In practice, nitrogen leakage into the vial does not necessarily result in bacterial contamination.

Leakage of water from the thawing bath into the vial, or water from the thawing bath adhering to the vial contaminating the culture on transfer from the vial, will invariably result in bacterial contamination. If there is obvious leakage into the vial, it should be discarded. It has been found to be advantageous to wipe vials with 70% ethanol prior to aseptic transfer of the culture into fresh medium.

Quality control

Considerable emphasis has been placed on the requirement for high levels of post-thaw viability, which are needed to reduce the possibility of selecting for a preservation tolerant subpopulation. High post-thaw viability also minimises the incubation time required to regrow a culture to a suitable density for distribution or use. At CCAP a minimum viability of 50% is required for non-clonal cultures before they are solely maintained in the cryopreserved state.

Vital staining, morphological appearance and motility can all be used as indices of viability (McLellan 1989; Fenwick and Day 1992). However, only cell division/ regrowth post-thaw can give an absolute value of viability. The following techniques are examples of cell regrowth viability tests.

6.2.6 Colony Formation in Agar

Viability tests

This method is particularly useful for *unicellular* strains.

1 ml aliquots of logarithmic dilutions (diluted in appropriate growth medium) of the thawed culture and control (untreated non-frozen culture) are pipetted into a sterile Petri dishes, to which growth medium containing agar (1% v/v) at 40 °C is added. Cells are then dispersed by gentle agitation and the Petri dish sealed using Parafilm inverted and incubated as detailed in Procedure, Section 6.1. After appropriate incubation, usually 7–28 days, depending on the growth rate of the alga being tested, colonies within the agar are counted.

For non-motile organisms spread plates of logarithmic dilutions may be used. This has not been found to be as satisfactory, because the ability to form discrete colonies may be related to the moisture level at the agar surface. It is therefore standard practice to use the pour-plate technique for both motile and non-motile organisms.

It is worth noting that many unicellular strains form aggregates, therefore colonies in pour plates can arise from either unicells or clumps of cells. When carrying out viability assays, it is vital that this cellular multiplicity be taken into account for both control and freeze/thawed cultures. The frequency of cell aggregation is determined microscopically from three samples of at least 500 cells. The actual viability can then be estimated as outlined below:

Multiplicity = No. of cells /no. of clumps
Total cell number = No. of colonies × multiplicity
Percentage viability = Total no. of cells before treatment /total no. of cells after treatment ×100.

6.2.7 Most Probable Number Technique

This method, based on those of Throndsen (1978) and Meyer (1985), is less satisfactory than the previous system, as it has a high intrinsic level of variability, which can easily be exacerbated by the incubation conditions, e.g. non-uniform illumination. It is used for organisms which will not form colonies in agar and filamentous cyanobacteria.

To each well in a Micro-well plate (Nunc), 0.35 ml of medium is added. Five replicates of a series of four 1/8 dilutions of freeze /thawed and control samples are then added to the wells and the plates are then covered with clear plastic, to minimise evaporation before being incubated for up to 14 days under the conditions outlined in Procedure, Section 6.1. After incubation, the wells are observed under a dissecting microscope and assessed for cell growth. The general formula for most probable number (Swaroop 1938) can then be used to calculate a table listing the MPN for any particular well survival combination (Appendix) and the post-thaw viability calculated on comparison of control and treated cultures.

6.2.8 Other Quality Control Factors

The quality control factors detailed in Section 6.1 are equally valid for thawed cryopreserved strains.

References

Anonymous (1991) Diarrhoeal diseases. Weekly epidemiological record – WHO 66:241–242

Ben Amotz A, Gilboa A (1980) Cryopreservation of marine unicellular algae II. Induction of freezing tolerance. Mar Ecol 2:157–161

Collins M (1978) Algal toxins. Microbiol Rev 42:725–746

Day JG (1992) Cryopreservation of microalgae and cyanobacteria. In: ICCC–VII Symposium Abstr. 12–16 Oct. 1992. Beijing, China, p 2

Day JG, Turner MF (1992) Algal culture collections and biotechnology. In: Watanabe MM (ed) Proceedings of the symposium on culture collections of algae NIES, Tsukuba, pp 11–28

Fenwick C, Day JG (1992) Cryopreservation of *Tetraselmis suecica* cultured under different nutients regimes. J Appl Phycol 4:105–109

Fetter BF, Klintworth GK, Nielsen HS (1971) Protothecosis – algal infection. In: Baker RD (ed) Human infection with fungi, actinomycetes and algae. Springer Berlin Heidelberg New York, pp 1081–1093

Friedman EI (1980) Endolithic microbial life in hot and cold deserts. Origins of life 10:223–235

Grout BWW, Westcott RJ, Henshaw GG (1978) Survival of shoot meristems of tomato seedlings frozen in liquid nitrogen. Cryobiology 15:478–483
Hawksworth DI, Mound LA (1991) Diversity data bases: The crucial significance of collections. In: Hawksworth DI (ed) The biodiversity of microorganisms and invertebrates: its role in sustainable agriculture. CAB International, Wallingford, pp 17–29
Leeson EA, Cann JP, Morris GJ (1984) Maintenance of algae and protozoa. In: Kirsop BE, Snell JJS (eds) Maintenance of microorganisms. Academic Press, London, pp 131–160
Malik KA (1992) A new method for the preservation of unicellular algae by drying. In: ICCC-VII Symposium Abstr 12–16 Oct. 1992. Beijing, China, p 46
McLellan MR (1989) Cryopreservation of diatoms. Diatom Res 4:301–318
Meyer MA (1985) Cryopreservation of a marine diatom. Thesis, Texas A & M University, Galveston
Morris GJ (1978) Cryopreservation of 250 strains of Chlorococcales by the method of two-step freezing. Br Phycol J 13:15–24
Morris GJ (1981) Cryopreservation. Institute of Terrestrial Ecology, Cambridge
Morris GJ, Clarke A, Fuller BJ (1980) Methanol as a cryoprotective addititive for *Chlorella.* Cryo Lett 1:121–128
Nerad TA (1991) ATCC catalogue of protists. ATCC, Rockville.
O'Donnell AG (1992) Chemical and molecular methods in the analysis of microbial diversity. In: ICCC-VII Abstr 12–16 Oct 1992. Beijing, China, p 9
Olsen JL (1990) Nucleic acids in algal systematics. J Phycol 26:209–214
Pearson BM, Jackman PJH, Painting KH, Morris GJ (1990) Stability of genetically manipulated yeasts under different cryopreservation regimes. Cryo Lett 11:205–210
Rai AN, Rowell P, Stewart WDP (1984) Evidence for the ammonium transport system in free-living and symbiotic cyanobacteria. Arch Microbiol 137:241–246
Reisser W (1985) The taxonomy of green algae endosymbiotic in ciliates and a sponge. Br Phycol J 19:309–318
Rodgers GA, Stewart WDP (1977) The cyanophyte-hepatic symbiosis. New Phytol 78:441–458
Starr RC, Zeikus JA (1987) UTEX-The culture collection of algae at the University of Texas at Austin. J Phycol 23:1–47
Stewart WDP, Rowell P, Rai AN (1983) Cyanobacteria-eukaryotic plant symbioses. Ann Microbiol 134B:205–228
Swaroop S (1938) Numerical estimation of *E. coli* by dilution method. Indian J Med Res 26:353–378
Thompson AS, Rhodes JC, Pettman I (1988) Culture collection of algae and protozoa, catalogue of strains. Culture collection of algae and protozoa, Ambleside
Throndsen J (1978) The dilution-culture method. In: Sournia A (ed) Phytoplankton manual. UNESCO, Paris, pp 218–224
Watanabe MM, Satake KN (1991) NIES-collection list of strains. NIES, Tsukuba
Wilcox LW, Wedemayer GJ (1985) Dinoflagellates with blue-green chloroplasts derived from an endosymbiotic eukaryote. Science 227:192–194

Appendix. Most Probable Number Table

Most probable number table calculated for four serial 1/8 dilutions with five replicates per dilution. Left hand and top margin indicate number of wells with growth per dilution row (i.e., 5503 would indicate 302 viable cells)

1 2 3[a]-Dilution row-4

	5	4	3	2	1	0
5 5 5	9000	6593	3784	2346	1587	1125
5 5 4	2101	1697	1363	1093	844	636
5 5 3	1267	1060	872	699	540	399
5 5 2	890	743	604	474	385	264
5 5 1	654	536	425	327	247	185
5 5 0	484	388	302	232	177	134
5 4 5	492	440	388	339	293	250
5 4 4	419	371	325	281	241	203
5 4 3	356	312	271	232	197	164
5 4 2	301	262	225	191	160	131
5 4 1	253	218	186	155	128	102
5 4 0	212	181	152	125	100	77
5 3 5	280	253	227	202	177	154
5 3 4	247	222	197	174	151	129
5 3 3	217	193	170	148	127	106
5 3 2	189	167	145	124	104	85
5 3 1	164	143	122	103	84	66
5 3 0	140	120	101	83	65	49
5 2 5	198	179	161	143	126	109
5 2 4	176	159	141	124	107	91
5 2 3	156	139	122	106	90	74
5 2 2	137	121	104	89	73	58
5 2 1	119	106	87	72	57	44
5 2 0	102	86	71	57	44	33
5 1 5	151	132	122	108	94	90
5 1 4	135	121	107	93	79	66
5 1 3	119	106	92	78	65	52
5 1 2	104	91	76	64	52	40
5 1 1	90	77	64	51	40	31
5 1 0	76	63	51	40	30	23
5 0 5	120	105	96	84	72	60
5 0 4	107	95	83	71	59	48
5 0 3	94	82	70	59	47	37
5 0 2	81	69	58	47	37	29
5 0 1	69	57	47	37	29	22
5 0 0	57	46	37	28	22	17

Most probable number (continued)

1 2 3[a]-Dilution row-4

	5	4	3	2	1	0
4 5 5	94	87	80	74	67	61
4 5 4	86	80	73	67	61	55
4 5 3	80	73	67	60	54	48
4 5 2	73	66	60	54	48	42
4 5 1	66	60	54	48	42	36
4 5 0	59	53	47	42	36	31
4 4 5	82	76	70	64	58	52
4 4 4	75	69	63	58	52	46
4 4 3	69	63	57	52	46	40
4 4 2	63	57	51	46	40	35
4 4 1	57	51	45	40	35	30
4 4 0	51	45	40	35	30	25
4 3 5	72	66	61	55	50	44
4 3 4	66	60	55	49	44	39
4 3 3	60	55	49	44	39	34
4 3 2	54	49	44	38	34	29
4 3 1	49	43	38	33	29	24
4 3 0	43	38	33	29	24	20
4 2 5	63	58	53	47	42	37
4 2 4	58	52	47	42	37	33
4 2 3	52	47	42	37	32	28
4 2 2	47	42	37	32	28	24
4 2 1	42	37	32	28	24	20
4 2 0	37	32	28	24	20	16
4 1 5	55	50	45	41	36	31
4 1 4	50	45	40	36	31	27
4 1 3	45	40	36	31	27	23
4 1 2	40	36	31	27	23	19
4 1 1	35	31	27	23	19	16
4 1 0	31	27	23	19	16	13
4 0 5	48	44	39	35	30	26
4 0 4	43	39	34	30	26	22
4 0 3	39	34	30	26	22	19
4 0 2	34	30	26	22	19	16
4 0 1	30	26	22	19	16	13
4 0 0	26	22	19	15	12	10
3 5 5	52	48	45	42	38	25
3 5 4	48	45	41	38	35	32
3 5 3	45	41	38	35	31	28

Most probable number (continued)

1 2 3[a]-Dilution row-4

	5	4	3	2	1	0
3 5 2	41	38	35	31	28	25
3 5 1	38	34	31	28	25	22
3 5 0	34	31	28	25	22	19
3 4 5	47	44	40	37	34	31
3 4 4	44	40	37	34	31	28
3 4 3	40	37	34	31	28	25
3 4 2	37	34	31	28	25	22
3 4 1	34	30	27	24	22	19
3 4 0	30	27	24	22	19	16
3 3 5	43	39	36	33	30	27
3 3 4	39	36	33	30	27	24
3 3 3	36	33	30	27	24	21
3 3 2	33	30	27	24	21	18
3 3 1	30	27	24	21	18	16
3 3 0	27	24	21	18	16	13
3 2 5	38	35	32	29	26	24
3 2 4	35	32	29	26	24	21
3 2 3	32	29	26	23	21	18
3 2 2	29	26	23	21	18	16
3 2 1	26	23	21	18	15	13
3 2 0	23	21	18	15	13	11
3 1 5	34	31	29	26	23	20
3 1 4	31	29	26	23	20	18
3 1 3	28	26	23	20	18	15
3 1 2	26	23	20	18	15	13
3 1 1	23	20	18	15	13	10
3 1 0	20	18	15	13	10	8
3 0 5	31	28	25	23	20	17
3 0 4	28	25	23	20	17	15
3 0 3	25	22	20	17	15	13
3 0 2	22	20	17	15	13	10
3 0 1	20	17	15	13	10	8
3 0 0	17	15	13	10	8	6
2 5 5	36	34	32	29	27	25
2 5 4	34	32	29	27	25	22
2 5 3	31	29	27	25	22	20
2 5 2	29	27	25	22	20	18
2 5 1	27	24	22	20	18	16
2 5 0	24	22	20	18	16	14

Most probable number (continued)

1 2 3[a]-Dilution row-4	5	4	3	2	1	0
2 4 5	33	31	29	26	24	22
2 4 4	31	29	26	24	22	20
2 4 3	29	26	24	22	20	18
2 4 2	26	24	22	20	18	16
2 4 1	24	22	20	18	15	13
2 4 0	22	20	18	15	13	11
2 3 5	30	28	26	24	22	19
2 3 4	28	26	24	22	19	17
2 3 3	26	24	22	19	17	15
2 3 2	24	22	19	17	15	13
2 3 1	21	19	17	15	13	11
2 3 0	19	17	15	13	11	9
2 2 5	28	26	23	21	19	17
2 2 4	25	23	21	19	17	15
2 2 3	23	21	19	17	15	13
2 2 2	21	19	17	15	13	11
2 2 1	19	17	15	13	11	9
2 2 0	17	15	13	11	9	7
2 1 5	25	23	21	19	17	15
2 1 4	23	21	19	17	15	13
2 1 3	21	19	17	15	13	11
2 1 2	19	17	15	13	11	9
2 1 1	17	15	13	11	9	7
2 1 0	15	13	11	9	7	5
2 0 5	23	21	19	17	15	13
2 0 4	21	19	17	15	13	11
2 0 3	18	17	15	13	11	9
2 0 1	15	13	11	9	7	5
2 0 0	13	11	9	7	5	4
1 5 5	28	26	24	22	21	19
1 5 4	26	24	22	21	19	17
1 5 3	24	22	21	19	17	15
1 5 2	22	21	19	17	15	14
1 5 1	21	19	17	15	14	12
1 5 0	19	17	15	13	12	10
1 4 5	26	24	22	20	19	17
1 4 4	24	22	20	19	17	15
1 4 3	22	20	19	17	15	13
1 4 2	20	19	17	15	13	12

Most probable number (continued)

1 2 3[a]-Dilution row-4

	5	4	3	2	1	0
1 4 1	18	17	15	13	12	10
1 4 0	17	15	13	12	10	8
1 3 5	24	22	20	18	17	15
1 3 4	22	20	18	17	15	13
1 3 3	20	18	17	15	13	11
1 3 2	18	17	15	13	11	10
1 3 1	17	15	13	11	10	8
1 3 0	15	13	11	10	8	6
1 2 5	22	20	18	16	15	13
1 2 4	20	18	16	15	13	11
1 2 3	18	16	15	13	11	10
1 2 2	16	15	13	11	10	8
1 2 1	15	13	11	10	8	6
1 2 0	13	11	10	8	6	5
1 1 5	20	18	16	15	13	11
1 1 4	18	16	15	13	11	10
1 1 3	16	14	13	11	10	8
1 1 2	14	13	11	10	8	6
1 1 1	13	11	10	8	6	5
1 1 0	11	10	8	6	5	3
1 0 5	18	16	14	13	11	9
1 0 4	16	14	13	11	9	8
1 0 3	14	13	11	9	8	6
1 0 2	13	11	9	8	6	5
1 0 1	11	9	8	6	5	3
1 0 0	9	8	6	5	3	2
0 5 5	22	21	19	18	16	15
0 5 4	21	19	18	16	15	13
0 5 3	19	18	16	15	13	12
0 5 2	18	16	15	13	12	10
0 5 1	16	15	13	12	10	9
0 5 0	15	13	12	10	9	7
0 4 5	21	19	18	16	15	13
0 4 4	19	18	16	15	13	12
0 4 3	18	16	15	13	12	10
0 4 2	16	15	13	12	10	9
0 4 1	15	13	12	10	9	7
0 4 0	13	12	10	9	7	6

Most probable number (continued)

1 2 3[a]-Dilution row-4	5	4	3	2	1	0
0 3 5	19	18	16	15	13	12
0 3 4	18	16	15	13	12	10
0 3 3	16	15	13	12	10	9
0 3 2	15	13	12	10	9	7
0 3 1	13	12	10	9	7	6
0 3 0	12	10	9	7	6	4
0 2 5	17	16	14	13	12	10
0 2 4	16	14	13	12	10	9
0 2 3	14	13	12	10	9	7
0 2 2	13	12	10	9	7	6
0 2 1	12	10	9	7	6	4
0 2 0	10	9	7	6	4	3
0 1 5	16	14	13	11	10	9
0 1 4	14	13	11	10	9	7
0 1 3	13	11	10	9	7	6
0 1 2	11	10	9	7	6	4
0 1 1	10	9	7	6	4	3
0 1 0	9	7	6	4	3	2
0 0 5	14	13	11	10	8	7
0 0 4	13	11	10	8	7	6
0 0 3	11	10	8	7	6	4
0 0 2	10	8	7	6	4	3
0 0 1	8	7	6	4	3	2
0 0 0	7	6	4	3	2	1

[a] Dilution row 1 = 1/8 dilution
Dilution row 2 = 1/64
Dilution row 3 = 1/512

List of Suppliers

Costar Corp., 205 Broadway, Cambridge, MA132140, USA.
Fryka Kaltetechnic, (UK Supplier) Camlab Ltd., Nuffield Rd., Cambridge, CB4 1TH, UK.
GallenKamp, Belton Road West, Loughborough. Leicestershire, LG11 0TR, UK.
NUNC Intermed, Post Box 280 Kamstrup, DK 4000 Roskilde, Denmark.
Oxoid: Unipath Ltd., Wade Rd. Basingstoke, Hampshire, RG240PW, UK.
Planer Products, Windmill Rd., Sudbury-on-Thames, Middlesex, TW16 7TH, UK.
Waterlife Research Industries Ltd., 476 Bath Rd., Longford, West Drayton, Middlesex, UB7 0ED, UK.

7 Cryopreservation by Vitrification

L.E. Towill

Cryogenic storage (cryopreservation) of shoot tips or buds is recognized as the major method for long-term germplasm storage for species that are maintained clonally (Withers 1988; Towill 1990a). Cryopreservation of protoplasts, cells, tissues, and somatic embryos is also needed for many tissue culture operations to assure line purity and performance, and to avoid costly, frequent tranfers. Low temperatures reduce maintenance requirements and should provide safe, long-term storage if properly administered. Storage is accomplished either in liquid nitrogen (LN; −196 °C), or in the vapor phase over LN (ca. −160 to −180 °C). At these temperatures, molecular motion is greatly reduced, and no liquid water phase exists.

Previous research, and methods developed therewith, emphasized slow cooling or two-step cooling techniques to attain cryopreservation. These methods have not been successful with all materials. An alternative method for attaining cryopreservation is through vitrification, the transformation of a liquid to a glass. There are increasing numbers of theoretical and practical studies examining this preservation stategy in animal and plant systems (Fahy et al. 1987; Hirsh 1987; Rall 1987; Fahy 1988; Steponkus et al. 1992). Vitrification requires application of suitable concentrations of compounds such that with cooling, usually rapid, the system forms a glass (i.e. the solution vitrifies). Information about physical aspects of the process are described in MacFarlane and Forsyth (1990) and MacFarlane et al. (1992). Differential scanning calorimetry is used to demonstrate that the solution in which the cells are bathed does indeed vitrify under favorable conditions, but it is difficult to prove that the cell itself vitrifies. Nevertheless, circumstantial evidence suggests that this happens under defined conditions.

Objectives

The advantages of vitrification are that it is a simple technique (not requiring an expensive apparatus), may avoid some damaging consequences of ice formation, and may be applicable to larger pieces of tissue than are usually used for two-step cooling methods. Survival after cryogenic exposure has been reported with several different vitrification procedures for protoplasts (Langis and Steponkus 1990, 1991), cells (Langis et al. 1989; Uragami et al. 1989; Sakai et al. 1990), somatic embryos (Uragami et al. 1989; L.E. Towill, unpubl.), and shoot tips (Towill 1990b; Niino et al. 1991, 1992a,b; Yamada et al. 1991; Steponkus et al. 1992; Towill and Jarret 1992).

The general strategy for vitrification is to expose the shoot tips (or protoplasts, cells, somatic embryos, etc.) to low concentrations of the vitrification solution such that the permeable components have sufficient time to penetrate (the loading phase). Next, the shoot tips are exposed to higher concentrations of the vitrification solution. Since water permeability is much greater than solute permeability, the shoot tip dehydrates and components that permeated, as well as the endogenous solutes, are concentrated. Shoot tips are then rapidly cooled to facilitate glass formation.

It has become apparent that loading may not be necessary for shoot tips from some species, and, indeed, may be detrimental (Steponkus et al. 1992). The cell contents alone, when sufficiently dehydrated, apparently vitrify when cooled at a rapid rate. For these shoot tips, preculture, often in a moderate osmotic solution, prior to dehydration is necessary to achieve survival after vitrification.

This strategy is also apparent in encapsulation methods in which shoot tips are encapsulated in alginate, cultured in sucrose solutions, dehydrated, and then subjected to rapid cooling. This method has induced survival in shoot tips from potato (Fabre and Dereuddre 1990), pear (Dereuddre et al. 1990), carnation (Tannoury et al. 1991), grape (Plessis et al. 1991), and sugarcane (Paulet et al. 1993). Ostensibly, encapsulated shoot tips vitrify under the conditions employed.

The following experiments demonstrate these methods of vitrification. As with any method, details often differ with species and with cell type and tissue complexity. Shoot tips are emphasized here since they are the propagules used for germplasm preservation of a clonal line.

Equipment and materials

Equipment:

- sterile air transfer hood
- dissecting microscope and light source
- forceps, with insulated tips
- scalpel/razor blade fragment
- liquid nitrogen Dewar, wide-mouth type
- semen straws, (0.25 or 0.5 ml)
- straw closures or heat-sealing devise for straws
- scissors
- glass test tubes: 12×75 mm
- safety glasses or mask
- Pasteur pipettes
- petri dishes
- timer

Biological Materials:

- Greenhouse, growth chamber or micropropagated plants can be used as stock materials.

- Carnation, potato, mint, strawberry, and raspberries are easy to establish as in vitro plants, and information is available on the culture of isolated shoot tips. Contamination problems are reduced if in vitro plants are used. Mint and potato can be obtained readily and placed into culture and, thus, are used for the experiments described. Both are easily micropropagated using either apical or nodal sections cultured on a growth medium devoid of growth regulators. The vitrification methods described have been used successfully for several species of mint and various cultivars of potato, as well as apple, *Prunus* sp., endod, papaya, strawberry, carnation, and sweet potato (Towill, unpubl.).

Solutions:

- *Growth Media.* Both liquid and solidified media are required for these experiments. The growth medium used for excised shoot tips of potato and mint contains Murashige and Skoog, 1962 (MS) minerals and vitamins, $100\,mg\,l^{-1}$ myo-inositol, 2% sucrose and, if solidified, 0.7% agar (pH 5.7). The growth regulators for mint are $0.5\,mg\,l^{-1}$ benzylaminopurine (BAP) and $0.1\,mg\,l^{-1}$ indolebutyric acid (IBA) and for potato are $0.01\,mg\,l^{-1}$ BAP, $0.001\,mg\,l^{-1}$ naphthalene acetic acid (NAA), and $5\,mg\,l^{-1}$ gibberelic acid (GA_3). These media facilitate shoot development with minimal callus formation. Media are autoclaved and are stored in small aliquots in a refrigerator.

- *Vitrification Solutions.* The following solutions are prepared in advance and stored at 4 °C for up to 4 weeks with no apparent loss in effectiveness.

  – Solutions needed for method A

  PVS (identified as "PVS2" by Sakai et al. 1990, 1991) solution contains (wt/wt basis) 30% glycerol, 15% dimethylsulfoxide (DMSO) and 15% ethylene glycol (EG). The aqueous weight percentage (i.e. 40%) added contains MS minerals, vitamins, and 0.4 M sucrose (no growth regulators). Thus, sucrose is approximately 0.15 M in the vitrification solution. This solution is termed here 100% PVS. Since this solution must be sterile, prepare the glycerol and EG portion at somewhat more concentrated levels in the growth medium, and autoclave. After cool-

Have ready

ing, add DMSO (DMSO is sterile in a new bottle) and bring to the final volume with growth medium (0.4 M sucrose). The 100% PVS solution is diluted with growth medium (0.4 M sucrose; no growth regulators) to form a 60% (v/v) PVS and a 20% (v/v) PVS solution.

- Solutions needed for method B

  Two solutions are prepared. The first contains 2 M EG + 20% (w/v) polyethylene glycol-8000 (PEG-8000) in distilled water. The second contains 80% (w/w) propylene glycol in distilled water. Both solutions can be either autoclaved or filter sterilized.

- Solutions needed for method C

  This method uses the formulation 50% EG, 15% sorbitol, and 6% bovine serum albumin (BSA) (w/w) as described by Steponkus et al. 1992.

- *Dilution Medium.* The dilution medium for removing vitrification solution components from shoot tips is composed of MS minerals and vitamins, 1.2M sucrose, and 100mg/l myo-inositol (pH 5.7).

- *Encapsulation Media.* A liquid medium is prepared containing 3% sodium alginate (medium viscocity) in a modified MS minerals and vitamins solution from which calcium is omitted. A second medium contains a modified MS minerals and vitamins solution with a final concentration of 100 mM $CaCl_2$. Both media contain sucrose at 0.3 M, but could be prepared with higher concentrations.

7.1 Vitrification of Shoot Tips

7.1.1 Excision

Procedure

Select stock plants that are healthy and sufficiently old such that axillary shoot tips have expanded but not elongated. A cold acclimation period for the stock plant is often beneficial for those species which have the ability to acclimate (Reed 1988). Excise axillary shoot tips under a dissecting microscope using a razor blade fragment, scalpel, or edge of a hypodermic needle. The isolated shoot tip should be between 0.5 and 1 mm in length and contains about two to four distinct leaf primordia. In some cases, the outer larger leaf primordia must be removed to trim the shoot tip to this size. Work quickly to minimize desiccation. Transfer isolated shoot tips into a test tube (12 × 75 mm) containing a small amount of growth medium without growth regulators. Place shoot tips randomly into a series of test tubes to minimize position effects of shoot tips along the length of the shoot axis.

Three or four shoot tips are placed in each test tube. Use four to eight replicate test tubes per treatment.

7.1.2 Culture of Excised Shoot Tips

Shoot tips are used for vitrification immediately after excission or cultured prior to use for 1 to 2 days in liquid medium with growth regulators and either low levels of cryoprotectants (i.e. 5% DMSO or 3% ethylene glycol) or elevated levels of slowly penetrating osmotica (ca. 0.3–0.7 M sucrose). Modifications of this culture phase are probably necessary for other species.

7.1.3 Application of the Vitrification Solution

- *Method A*. Drain the test tubes containing the shoot tips with a Pasteur pipette and add approximately 0.25 ml of 20% PVS at room temperature. Swirl the tubes to mix the solution with any residual solution left in the test tube. After 60 min, drain this solution and add 0.25 ml of 60% PVS. Place the test tubes in an ice bath and incubate for 10 min. Remove this solution, and about 0.25 ml of 100% PVS solution (ice cold), and swirl. After about 5 min in this solution, transfer the shoot tips and a portion of the PVS solution from one test tube using a Pasteur pipette to a semen straw that was previously heat-sealed at one end. Add the solution such that there is no air pocket at the sealed end. Heat seal the other end of the straw (usually a small air pocket remains). The total time in the 100% PVS solution is 10 min prior to exposure to low temperature (or to dilution in the case of control exposed to just the vitrification solution). Practice transferring shoot tips with liquid to straws to assure that you can load all the straws within the 5-min time frame. Be careful not to draw shoot tips too far into the Pasteur pipette during transfer or they may adhere to the glass and be difficult to expel. Straws can also be sealed with cotton or small polypropylene closures supplied by some manufacturers.

- *Method B*. In this method, drain medium from test tubes containing the shoot tips and add 0.25 ml of distilled water. Then at 15-min intervals add four additions of 0.0625 ml of 2 M ethylene glycol and 20% PEG-800. At 60 min from the first addition, transfer test tubes to an ice bath and add 0.5 ml of 80% propylene glycol. Shoot tips remain at 0 °C in this medium for 15 min. During the final 5 min of this 15-min incubation, transfer shoot tips with liquid to straws as described previously.

- *Method C.* Shoot tips are cultured within 0.4–0.7 M sucrose for 1–2 days. No "loading" phase is used. Add the EG/sorbitol/BSA vitrification solution and incubate for 20 min at room temperature (potato) or for 15 min at 0 °C (carnation). Other species will require some modification of preculture duration and exposure to vitrification solution.

7.1.4 Cooling to Low Temperatures

Place semen straws, loaded with shoot tips, and solution, sealed, at about −160 °C and hold for 30 min. This temperature is approximated by filling a wide-mouth Dewar about one third full of LN and suspending a screen above the liquid level. This step reduces the incidence of glass fracturing within the straw during subsequent transfer to LN. Keep the Dewar loosely covered when not manipulating the straws. After 30 min in the vapor phase, immerse the straws directly into LN. Alternatively, if a LN refrigerator is available, store the straws directly in the vapor phase. **Take Care** *Be sure to wear safety glasses or a protective mask when using Dewars and working with LN (filling with LN, immersing materials, carrying, etc.) in case of an implosion of the Dewar.*

7.1.5 Warming

Remove straws from LN or vapor phase storage using precooled insulated forceps. Hold the straw by the end in room temperature air for 10 s prior to immersing it in sterile water at about 22 °C. Leave straws in water for about 1 min. This is long enough for the solution to become liquid within the straw and for the temperature to rise to about 0 °C. Transfer from LN *directly* into water often causes excessive glass cracking, which can reduce viability, perhaps due to devitrification (Williams and Carnahan 1990). Occasionally, cracking does not occur, but devitrification does (the contents of the straw become white).

7.1.6 Dilution

Cut the end of the straw with sterile scissors and expel the contents into 3 ml of 22 °C dilution medium. You may have to flush the straw with diluent using a Pasteur pipette if shoot tips adhere to the straw. Remove shoot tips from diluent after 30 min and place them on culture medium. The concentration of the diluent may need to be altered to improve survival for some species.

7.1.7 Culture

Poured melted growth medium containing growth regulators and agar into petri dishes and allow to solidify. Transfer shoot tips from dilution medium to the agar surface using a Pasteur pipette. Remove all excess liquid from around the shoot tip. Be careful not to embed the shoot tip within the agar. Wrap petri dishes with Parafilm to retard moisture loss and store dishes under low light conditions. To improve growth, transfer shoot tips at regular intervals (1–3 weeks) to plates with fresh medium.

7.1.8 Viability

Make growth observations using a dissecting microscope periodically over a 4- to 6-week period. Shoot tip expansion in controls often begins within 3 days, but vitrified shoot tips develop more slowly, often after 7 days or more. Growth rate of shoot tips that show some signs of growth or which remain green can be enhanced by transferring to fresh medium.

7.2 Encapsulation/Dehydration Method

7.2.1 Excision and Encapsulation

Excise shoot tips and briefly hold in growth medium as described in Section 7.1. Drain shoot tips, rinse, and resuspend in the alginate solution. Form beads by dropping a small volume of the alginate medium containing a shoot tip into the 100 mM calcium medium (within a beaker). The size of the drop can be adjusted by using pipettes of different bores. Three- to 5-mm diameter beads are desired. Retain beads in the calcium medium for 15–30 min to allow them to set completely.

7.2.2 Culture and Dehydration

Place alginate beads into a flask containing MS medium with 0.5 M sucrose and shake overnight. The next day, transfer beads to MS medium with 0.7 M sucrose, and again shake overnight. The following day, blot the beads to remove excess moisture and place in an uncovered petri dish within a laminar flow hood. Dry the beads in the air flow for about 3–4 h. Because drying rates vary with humidity, it is desirable in a separate experiment to measure bead moisture content with time of drying. Moisture content is determined gravimetrically.

7.2.3 Cooling and Warming

Use beads that have been dried to a moisture content of about 15–20% (wet weight basis). Place three to five beads in a sterile cryoampoule and immerse into liquid nitrogen. After a minimum of 1 h in the cryogen, retrieve vials and thaw in sterile warm water (2–4 min).

7.2.4 Dilution and Culture

Remove beads from the cryoampoule and place on a growth medium with a somewhat elevated sucrose level (0.2 M sucrose). Culture in dim light for 2–4 days. Then place the bead on medium with the usual sucrose content (ca. 2%). Culture the bead so that the shoot tip is close to the agar. After about 1 week, remove the shoot tip from the bead and culture on fresh growth medium. For some species, the shoot tip does not need to be removed from the bead.

7.3 Vitrification of Cells and Protoplasts

7.3.1 Cells

- *General Method.* The following was used by Sakai et al. 1991 for citrus nucellar cells. Use linear growth phase cells. Centrifuge or allow cells to settle, discard supernatant and add PVS vitrification solution to cells. Total exposure time in this solution is 5 min at 25 °C. Place samples into semen, straws, heat-seal the ends, and cool in LN as described above. Thawing is also as described above. Cut straws and expel cells into a 1.2-M sucrose dilution medium at 25 °C. After 10 min, harvest cells, and resuspend in growth medium. Determine viability on an aliquot of the cells with the fluorescein diacetate test (FDA) (Widholm 1972) or tetrazolium test (Towill and Mazur 1975). For growth estimates, discard most of the supernatant, and place cells on two pieces of filter paper. Place the two pieces on a solidified growth medium for several hours. Transfer the top filter paper containing cells to fresh, solidified growth medium, and culture for several days to observe growth.

- *Modifications*. Since there are only a few articles describing vitrification of plant cells, modifications of the general method described above will be needed for different species. For some lines, greater survival occurs if stock cell suspensions are first cultured for several days in a medium containing elevated sucrose levels (Takano and Tamura 1992). For some species, it is beneficial to expose cells briefly to a permeating cryoprotectant (termed the loading step) prior to application of the vitrification solution. Langis et al. (1989) used 1.5 M EG in growth medium for 90 min at 0 °C for loading.

7.3.2 Protoplasts

This method is reported by Langis and Steponkus (1991) for non-cold-acclimated rye leaf protoplasts. After isolation, protoplasts are held in an isotonic medium containing 0.53 osmolal sorbitol, 1 mM $CaCl_2$, 1 mM MES and 1% BSA. Mix a small volume of this protoplast suspension with a large volume of the isotonic medium containing 1.5 M EG and hold for 20 min at 20 °C. Centrifuge to pellet the protoplasts and discard the supernatant. Cool the pellet to 2 °C and add a chilled vitrification solution containing 24% EG, 30% sorbitol and 6% BSA (weight percentages). Suspend well and dispense into straws. The time in the vitrification solution prior to cooling to LN is 1 min. Cooling and warming are as described above. Straws are cut and the contents expelled into a 1.3 M sorbitol solution containing 1 mM $CaCl_2$, 1 mM MES (pH 6) and 0.2% BSA. Assess viability with the FDA method.

7.4 Further Experiments

For each species, the experimenter should initially determine what percentage of shoot tips survive excision and culture. Once skill in handling the small shoot tips is achieved, address the issues described below. The general procedure may need to be adjusted to increase survival for different species or lines. Among variables that may need to be examined are time and temperature of exposure to a given level of vitrification solution, diluent concentration, and time in diluent.

7.4.1 Toxicity

Vary the time that shoot tips are exposed to different concentrations of the vitrification solutions of methods A, B, or C prior to dilution. Determine the effects of exposure time on percentate survival and mode of shoot-tip development. Does callus form? Does the shoot tip directly develop into a "normal" shoot? Do leaves develop but no shoots form? In these experiments, be certain to include controls that are not exposed to any vitrification solutions and controls that are exposed only to the dilution solution. Perform a time course exposure (e.g. 10, 20, 30, and 40 min) of shoot tips to only the concentrated vitrification solution (e.g. 100% PVS) to determine the effects of desiccation on survival and mode of growth.

7.4.2 Cryopreservation

Use the longest exposure time that did not adversely affect survival (as determined in Sect. 7.2.1) and expose the shoot tips to LN and warm as described (Sects. 7.1.4–7.1.5). Determine whether it is advantageous to expose the shoot tips to dilute solutions of permeating cryoprotectants for periods of time prior to exposure to the concentrated solution and to LN. Be sure to include appropriate control straws for all treatments. Always include controls that have not been treated in any manner other than excission and culture to determine whether these processes are damaging.

7.4.3 Encapsulation/Dehydration

Since preculture is important, determine the concentration of sucrose and duration of exposure necessary to optimize survival. Should the exposure to sucrose be gradual? Also examine the effect of time of desiccation on survival, both for control and LN-exposed samples.

Comments To obtain *survival* with vitrification, cells are first loaded with sufficient amounts of glass-forming compounds such that glasses can be formed. Cells are then exposed to high levels of compounds to osmotically concentrate the internal contents further. In some cases, desiccation alone, without loading, may suffice to concentrate endogenous compounds and allow for survival after low temperature exposure. High concentrations of cryoprotectants may be toxic, due either to osmotic excursions or direct effects on biochemical processes (Fahy 1986; Arakawa et al. 1990). Since toxicity has a kinetic component, exposure times and temperatures are important. Some compounds may be added to the vitrification solution to alleviate toxicity (Fahy et al. 1990); however, this has not been adequately examined in plant systems. Of course, multicellular systems add complexities for solute and water permeations and toxicity.

Cooling must be sufficiently fast to facilitate glass formation, since the concentrations of cryoprotectants added and those which develop within cells are not ideal. Critical cooling rate information is available for some vitrification solutions (Sutton 1991), but not for those used in plant systems.

Warming rate also influences survival. If too slow, devitrification has sufficient time to occur and this decreases survival. Again, critical warming rates are not defined for most vitrification solutions (Boutron and Mehl 1990). In practice, rates are selected that are sufficiently rapid to avoid loss in viability and yet not fracture the external glass. The stability and characteristics of a glass with mode of formation and length of storage are largely unknown (Chang and Baust 1991). The methods described above contain many variables (for example: preculture time and conditions, content of the vitrification solution, time in various concentrations of the vitrification solution, concentration of diluent, time in diluent, temperature of exposure to vitrification solutions and to diluent). Optima for these variables may differ among species and, thus, are usually determined empirically. Many physical, chemical, and biological questions remain, answers to which would enhance development and applicability of the technique.

The physiological status of the stock plant, reflected in the biochemical status of the excised shoot tip, affects survival and performance after cryopreservation. Few studies critically address this important issue.

In summary, vitrification is a relatively new method for cryopreservation of plant germplasm. Undoubtedly, knowledge about permeation kinetics, stabilities of glasses, improvements in the constituents of vitrification solutions, and the physiological status of the plant will provide new or modified cryopreservation methods that are suitable for a range of species or of clones within a species.

References

Arakawa T, Carpenter J, Kita Y, Crowe J (1990) The basis of toxicity of certain cryoprotectants: a hypothesis. Cryobiology 27:401–415

Boutron P, Mehl P (1990) Theoretical prediction of devitrification tendency: determination of critical warming rates without using finite expansions. Cryobiology 27:359–377

Chang ZH, Baust JG (1991) Physical aging of glassy state: DSC study of vitrified glycerol systems. Cryobiology 28:87–95

Dereuddre J, Scottez C, Arnaud Y, Duron M (1990) Resistance of alginate-coated axillary shoot tips of pear tree (*Pyrus communis* L. cv. Beurre Hardy) in vitro plantlets to dehydration and subsequent freezing in liquid nitrogen: effects of previous cold hardening. CR Acad Sci Paris 310:317–323

Fabre J, Dereuddre J (1990) Encapsulation-dehydration: a new approach to cryopreservation of *Solanum* shoot tips. Cryo Lett 11:413–426

Fahy GM (1986) The relevance of cryoprotectant "toxicity" to cryobiology. Cryobiology 23:1–13

Fahy GM (1988) Vitrification. In: McGrath JJ, Diller KR (eds) Low temperature biotechnology: emerging applications and engineering contributions. Am Soc Mechanical Engineers, New York, pp 113–146

Fahy GM, Levy DI, Ali SE (1987) Some emerging principles underlying the physical properties, biological actions, and utility of vitrification solutions. Cryobiology 24:196–213

Fahy GM, Lilley TH, Linsdell H, Douglas MSJ, Meryman HT (1990) Cryoprotectant toxicity and cryoprotectant toxicity reduction: in search of molecular mechanisms. Cryobiology 27:247–268

Hirsh AG (1987) Vitrification in plants as a natural form of cryoprotection. Cryobiology 24:214–218

Langis R, Steponkus PL (1990) Cryopreservation of rye protoplasts by vitrification. Plant Physiol 92:666–671

Langis R, Steponkus PL (1991) Vitrification of isolated rye protoplasts: protection against dehydration injury by ethylene glycol. Cryo Lett 12:107–112

Langis R, Schnabel B, Earle ED, Steponkus PL (1989) Cryopreservation of *Brassica campestris* L. cell suspensions by vitrification. Cryo Lett 10:421–428

MacFarlane DR, Forsyth M (1990) Recent insights on the role of cryoprotective agents in vitrification. Cryobiology 27:345–358

MacFarlane DR, Forsyth M, Barton CA (1992) Vitrification and devitrification in cryopreservation. In: Steponkus PL (ed) Advances in low-temperature biology, vol 1. JAI Press Ltd, London, pp 221–278

Moriguchi T, Akihama T, Kozaki I (1985) Freeze-preservation of dormant pear shoot apices. Jpn J Breed 35:196–199

Murashige T, Skoog F (1962) A revised medium for rapid growth and bio-assays with tobacco tissue cultures. Physiol Plant 15:473–497

Niino T, Sakai A, Matsumura T, Higuchi S (1991) Cryopreservation of apical meristems of white clover (*Trifolium repens* L.) by vitrification. Plant Sci 78:81–87

Niino T, Sakai A, Enomoto S, Magosi J, Kato S (1992a) Cryopreservation of invitro-grown shoot tips of mulberry by vitrification. Cryo Lett 13:303–312

Niino T, Sakai A, Yakuwa H, Nojiri K (1992b) Cryopreservation of in vitro-grown shoot tips of apple and pear by vitrification. Plant Cell Tissue Organ Cult 28:261–266

Paulet F, Engelmann F, Glaszmann JC (1993) Cryopreservation of apices of in vitro plantlets of sugarcane (*Saccharum* sp. hydrids) using encapsulation/dehydration. Plant Cell Rep 12:525–529

Plessis P, Leddet C, Dereuddre J (1991) Resistance to dehydration and to freezing in liquid nitrogen of alginate-coated shoot tips of grape vine (*Vitis vinifera* L. cv. Chardonnay). CR Acad Sci Paris 313:373–380

Rall WF (1987) Factors affecting the survival of mouse embryos cryopreserved by vitrification. Cryobiology 24:387–402

Reed BM (1988) Cold acclimation as a method to improve survival of cryopreserved *Rubus* meristems. Cryo Lett 9:166–171

Sakai A, Kobayashi S, Oiyama I (1990) Cryopreservation of nucellar cells of navel orange (*Citrus sinensis* Osb. var. *brasiliensis* Tanaka) by vitrification. Plant Cell Rep 9:30–35

Sakai A, Kobayashi S, Oiyama I (1991) Survival by vitrification of nucellar cells of Navel Orange (*Citrus sinensis* var. *brasiliensis* Tanaka) cooled to −196 °C. J Plant Physiol 137:465–470

Steponkus PL, Langis R, Fujikawa S (1992) Cryopreservation of plant tissues by vitrification. In: Steponkus PL (ed) Advances in low-temperature biology, vol 1. JAI Press Ltd, London, pp 1–62

Sutton RL (1991) Critical cooling rates to avoid ice crystallization in solutions of cryoprotective agents. J Chem Soc Faraday Trans 87:101–105

Takano M, Tamura S (1992) Vitrification of cultured cells of *Nicotiana tabacum*. Cryobiology 29:739

Tannoury M, Ralambosoa J, Kaminski M, Dereuddre J (1991) Cryoconservation by vitrification of alginate-coated carnation (*Dianthus caryophyllus* L) shoot tips of in vitro plantlets. CR Acad Sci Paris 313:633–638

Towill LE (1990a) Cryopreservation. In: Dodds JH (ed) In vitro methods for conservation of plant genetic resources. Chapman and Hall, London, pp 41–70

Towill LE (1990b) Cryopreservation of isolated mint shoot tips by vitrification. Plant Cell Rep 9:178–180

Towill LE, Jarret RL (1992) Cryopreservation of sweet potato [*Ipomoea batatas* (L.) Lam] shoot tips by vitrification. Plant Cell Rep 11:175–178

Towill LE, Mazur P (1975) Studies on the reduction of 2,3,5 triphenyltetrazolium chloride as a viability assay for plant tissue cultures. Can J Bot 53:1097–1102

Uragami A, Sakai A, Nagai M, and Takahashi T (1989) Survival of cultured cells and somatic embryos of *Asparagus officinalis* cryopreserved by vitrification. Plant Cell Rep 8:418–421

Widholm JM (1972) The use of flurescein diacetate and phenosafranine for determining viability of cultured plant cells. Stain Technol 47:189–194

Williams RJ, Carnahan DL (1990) Fracture faces and other interfaces as ice nucleation sites. Cryobiology 27:479–482

Withers L (1988) Germplasm preservation. In: Application of plant cell and tissue culture (Ciba Foundation Symposium 137) Wiley, Chicester, pp 163–177

Yamada T, Sakai A, Matsumura T, Higuchi S (1991) Cryopreservation of apical meristems of white clover (*Trifolium repens* L.) by vitrification. Plant Sci 78:81–87

8 Biochemical and Molecular Methods for Assessing Damage, Recovery and Stability in Cryopreserved Plant Germplasm

K. Harding and E.E. Benson

The reasons for using ultra-low-temperature storage techniques in the conservation of plant cells and organs are very varied. Because of these wide-ranging applications, there already exist many approaches to monitoring recovery and stability in cryopreserved plant tissues. However, it is important to distinguish two categories of assessment:

- *General.* This includes methods which are used in other areas of plant biotechnology and breeding. Their use in cryobiology is predetermined by the storage application (e.g. Benson and Hamill 1991). Examples include evaluating the effects of freezing on secondary product synthesis, foreign DNA (transformation) stability, chromosome stability (Ward et al. 1993), and the maintenance of morphogenic potential, and protoplast regeneration characteristics. Specific plant breeding methodology may also be utilised as a means of assessing post-freeze stability (field performance trials, phenotype studies, isoenzyme profiles). The authors consider that these approaches are not only beyond the scope of this chapter, but are also well documented in individual publications and other methodological texts.
- *Specialised.* The objectives of this chapter are to present a series of biochemical and molecular methods which provide a more specialised approach to evaluating damage, recovery and stability in cryopreserved plant tissues. These methods may be applied to both cells and organs and their use is especially encouraged to:
  - Aid basic research.
  - Increase our understanding of freezing injury and post-freeze molecular stability.
  - Permit the development of new storage strategies (e.g. for problematic tissues).

In many cases, the methods presented in this chapter have been adapted from other areas of cryobiology (the frozen food industry, mammalian organ storage) and plant molecular biology. All the methods have either been routinely used by the authors or are currently being applied to investigate stress in both frozen and unfrozen plant tissue cultures (e.g., for assessments of in vitro storage under normal and slow growth conditions). This experience enables the provision of supplementary "tips"; however, the methods are by no means exhaustive, and readers are directed to additional texts which may advance their further use. This is especially pertinent to the molecular techniques, which, to date, have been little exploited in the field of plant cryopreservation. Finally, emphasis has been placed on methods which do not necessarily require expensive analytical equipment, as these approaches are more amenable to working gene banks which may not have ready access to advanced instrumentation.

A Biochemical Methods

The use of biochemical tests to evaluate low-temperature injury is a common practice in the areas of cold hardening, seed storage, desiccation and environmental freeze tolerance. In selecting a strategy for investigating low-temperature-induced biochemical lesions, it is important to recognise that both freezing injury and recovery from that injury are very complex (Benson 1990). The evaluation of freezing damage must also be accompanied by viability and re-growth assessments. Often, it is better to use several complementary approaches to investigate damage and recovery. Many biochemical methods are themselves destructive, which may make their interpretation difficult when applied to badly damaged tissues. The concomitant assessment of unfrozen control tissue is therefore always recommended and non-destructive methods are especially useful. The cryopreservation of plant cells and organs often involves a pre- and post-freeze period of tissue culture. It is therefore important that the biochemical tests are applied both to cryopreservation and to associated in vitro manipulations.

8.1 Non-Destructive Volatile Hydrocarbon Monitoring of Freezing Damage and Recovery (Benson and Withers 1987)

Tissues are placed in gas-tight sampling vessels which permit the accumulation of "marker" volatile compounds; these are then withdrawn by a syringe and measured by gas chromatography. The use of non-destructive methods to assess freezing injury and recovery may be particularly useful. Damage can be evaluated in the same tissues over an extended time course and can

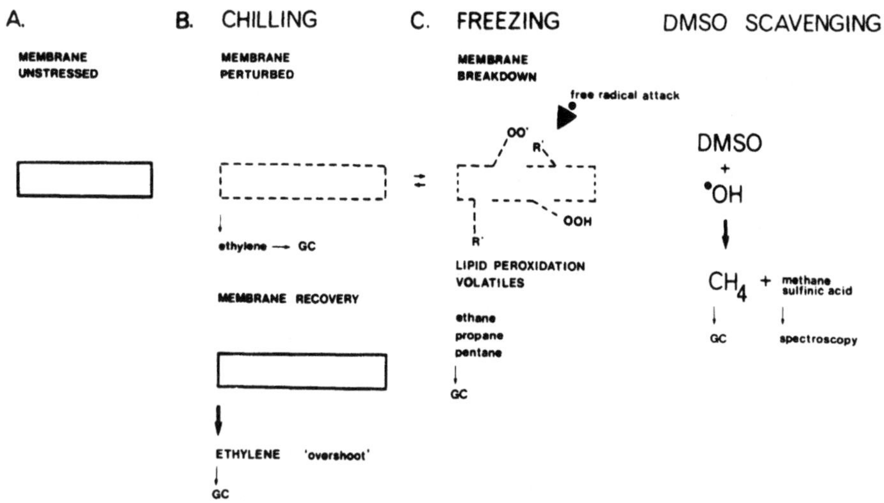

Fig. 1A–C. Temperature-induced (**B**–**C**) alterations in membrane conformation disturb the pathway involved in ethylene evolution (Field 1981). On recovery, the evolution of ethylene may be resumed after a lag phase. This often gives rise to an "overshoot" (**B**) in production (Field 1981, 1984; Corbineau et al. 1990). Ethane propane and pentane may be used as indicators of freezing stress (Harber and Fuchigami 1986). These volatiles are breakdown products of membranes (**C**) which have undergone free radical attack and lipid peroxidation (Dillard and Tappel 1979; Frank et al. 1978). In low-temperature-treated mammalian tissues the transfer back to normal temperatures has been associated with cold ischemic oxidative stress (Fuller and Green 1986). Similar changes have been observed in cryopreserved plant tissues (Benson and Withers 1987; Benson and Noronha-Dutra 1988). DMSO is a potent scavenger of the very toxic hydroxyl free radical (•OH) which is produced during oxidative damage. On reacting with •OH, DMSO forms several reaction products one of which is the methyl radical, CH_3. This readily abstracts H to form methane (CH_4) which can be also detected in headspace samples (Hebbel et al. 1982)

be directly correlated with recovery events. These marker volatiles are shown in Fig. 1.

Objective

- The purpose of this technique is to construct a profile of hydrocarbon "marker" volatiles (for stress, membrane damage, free radical activity and recovery) during the different stages of tissue culture, pre-growth, cryoprotection, freezing, thawing and recovery.

Equipment and materials

- a packed-column gas chromatograph (GC) fitted with a 2-m Poropak Q column connected to a flame ionisation detector (FID)
- a temperature programming facility (optional)
- supply of high purity carrier (nitrogen) and combustion (air and hydrogen) gases

- a chart recorder and/or data handling unit
- air-tight sampling syringes (1 and 5 ml)
- autoclavable sampling vials (1.5–12.0 ml, for small tissue samples) fitted with open top caps, sealed with inert silicone/Teflon disks
- high-quality rubber seals to fit conical culture flasks for cell suspensions
- ultra-pure standard gas mixtures for GC calibration

Preparation

8.1.1 Sampling Vessels

To permit accurate headspace sampling, the cultures under study have to be transferred to air-tight vessels containing the appropriate culture medium/cryoprotectant. For small quantities of tissue (e.g. shoot tips, embryos, callus) glass vials supplied by Pierce (see suppliers' list) are recommended. These are autoclavable and are supplied with inert silicone/Teflon disks which readily re-seal after syringe sampling, and they do not produce interfering volatiles. High-quality Suba seals are recommended for air-tight closure of conical culture flasks. However, it is most important to ensure that the sampling vessels and their closures do not produce ethylene (a common by-product of the manufacturing polymerisation process used for the production of seals). Many "rubber" seals produce ethylene and other volatile emissions which may interfere with the assay or be harmful to the plant tissues.

1. Use 1–4 ml glass vials for small pieces of tissue (shoot tips, lumps of callus, embryos). Conical flasks may be used for cell suspension cultures.
2. Record accurately the volume of the vessel by immersing in water and calculating the weight difference when empty.
3. Autoclave the vials, flasks, closures in the absence of plastics (e.g. autoclave bags). On removal from the autoclave, immediately aerate in a sterile flow hood to remove any contaminating volatiles.
4. Add accurate, known volumes of pre-growth additive, cryoprotectant, recovery media. Aerate to remove excess water.
5. Re-weigh and make any additional volume corrections for the remaining available headspace.
6. Add plant material, taking care not to damage the tissues and thus avoid stress volatile production.

7. Re-weigh the vessels and calculate the weight of plant material by subtraction.
8. Record the time at which the vessels were closed.

8.1.2 Gas Chromatograph

Full details of how to operate individual GCs are provided by manufacturers' manuals. The following are useful points for the present application. The levels of volatiles produced by cryopreserved plant tissues are very low. This may be particularly the case when using very small quantities of tissues (e.g. shoot tips). It is most important that the column is well conditioned ("cleaned") and detection by the FID is optimised according to manufacturers' instructions. These precautions will allow the detector to operate at maximum sensitivity without baseline noise.

Column Conditioning. Perform this as recommended by the manufacturer, in brief:

1. Disconnect the column from the FID.
2. Purge the column (10–15 $ml\,min^{-1}$ carrier gas) and in steps, gradually increase the oven temperature to final temperature of 150 °C (or as recommended).
3. Maintain at maximum temperature overnight.
4. Re-connect to the FID and check for baseline noise; if stable, the column is ready for use.

Routine Column Cleaning. Plant tissues produce a range of carbonyl volatiles which are retained on the column under the low oven temperatures used in the isothermal method. It is recommended that the column is purged of these volatiles every 15–20 injections. This may be done by increasing the column temperature to 50–100 °C for 10–20 min. Failure to do this may result in the appearance of "ghost peaks" which interfere with the routine chromatography. Carbonyl volatiles are often produced by cell suspension and callus cultures, but are less common in organised tissues.

GC Parameters. The following methods were established using a Perkin Elmer 8310 GC; however, they may also be applied to other GCs.

Method A. Isothermal programming for the separation of C1–C2 hydrocarbons.

Set the GC parameters to the following settings:

Oven temperature	60 °C
Isothermal time	3 min
Injection temperature	200 °C
Detector temperature	225 °C
Carrier gas flow	25 ml.min^{-1}
Attenuation range	1–4

Combustion gas supplies are set to manufacturers' recommendations.

Method B. Temperature programming to separate C1–C5 hydrocarbons.

Set gas parameters, attenuation, injection and detector temperatures as above.

Oven temperature 1	60 °C
Isothermal time 1	5 min
Ramp rate	30 °C.min^{-1}
Oven temperature 2	150 °C
Isothermal time 2	15 min

GC Calibration

1. Inject a known volume (concentration) of volatile standard into the GC. The instrument should be calibrated using the same factor of sensitivity (attenuation) as will be used for the samples. For example, do not calibrate with standards at an attenuation of 128 if plant samples require an attenuation of 1–2 for detection.
2. Record the peak retention times and peak heights/areas using a chart recorder. Alternatively, newer instruments have a data handling facility which may be programmed to identify and integrate peak data.

8.1.3 Headspace Sampling and Volatile Measurement

1. Withdraw headspace samples (0.5–2.0 ml) from the culture vessels using a gas tight syringe. In order to ensure complete mixing, the syringe is flushed out several times before removal.
2. Inject onto the column.
3. In between sampling different vessels, flush out the syringe several times in a sterile flow hood.

4. After sampling, expose the vessels to the air in a flow hood and aerate before re-sealing.

5. Record the time of re-sealing.

For sequential sampling, cultures are maintained in the vessels, and by using a range of different sampling containers, it is possible to construct a time course of volatile production during the different stages of cryopreservation and associated tissue culture techniques (pre-growth, cryoprotection, chilling, freezing, thawing, recovery).

DMSO concentrations used in cryoprotection are within the same range as those used for the hydroxyl radical scavenging assays (5–10% v/v). In the case of tissues frozen in cryoampoules, it is also possible to take headspace directly from the vials (Benson and Withers 1987). However, it is important to flush out plasticiser volatiles from the cryovials before use in the experiment.

Controls

A control vessel is always prepared which contains medium, but no plant tissue. This is important for DMSO/methane studies, as methane is an atmospheric component. Ambient methane levels are therefore subtracted from those of the samples. Previous studies have shown that the levels of methane accumulating in DMSO-treated carrot cell suspensions are considerably higher than vessel/culture controls or controls containing non-DMSO-treated plant tissue (Benson and Withers 1987).

Recommendation

The limits of detection of volatiles in headspace samples is dependent on the amount of plant material placed in the sampling vessels (shoot tips, cells, callus). On a fresh weight basis this can range from 0.05–2.0 g. In the case of shoot tips which are small but metabolically very active, ethylene (ethene) may be detected in a sample of five shoot tips. This is achieved if the head space is concentrated into a small sampling volume (1–2 ml) and the total headspace is withdrawn. However, to prevent tissue anaerobis, the vials must be aerated directly after sampling.

Results

Using a 2-m Poropak Q column and the isothermal method 1, volatiles are separated within 3 min in the sequence: methane, ethylene, ethane. For method 2, methane, ethylene and ethane separate as described in the isothermal method. During temperature programming, propane and pentane are eluted at approximately 6 and 14 min, respectively. Volatile evolution is usually expressed as a rate ppm or ng or $\mu g\,min^{-1}\,g^{-1}$ fresh weight.

In some plant tissues (Benson and Withers 1987), a major peak is eluted at 8 min, which is thought to be a C_4 hydrocarbon (iso-butane) or a composite peak of butane and butene (Frank et al. 1978). Ethylene, ethane and pentane have been used as the major indicators of lipid peroxidation in both plant and animal tissues and are readily identified using standard calibration gases. However, using a temperature programming facility, a range of other volatiles (acetaldehyde, hexanal, hexanol, ethanol, unsaturated C_4 hydrocarbons) may be detected. Many of these have been identified as "markers" of lipid peroxidation (Frank et al. 1978; Dillard and Tappel 1979). The complete separation and identification of this wide range of components may not be achieved by using a single chromatography technique. Unequivocal identification, by comparing separations on a second column (e.g. Porasil C, as recommended by Frank et al. 1978) and confirmation by GC-mass spectroscopy is strongly advised.

Important notes The exact retention times of the volatiles will be dependent on sample injection, column age and conditioning. However, it is important to ensure that ethylene and ethane are adequately separated; if they peak too closely, drop the oven temperature by 5–10 °C and/or reduce the carrier gas flow. Compare ethylene and ethane separations in known standards as co-elution of these gases, as one peak will produce erroneous results. On free radical scavenging, DMSO forms several other breakdown products in addition to methane. Thus, formaldehyde (Hallinan et al. 1991) and methane sulfinic acid (Babbs et al. 1989) have been used as markers for OH/DMSO in hepatocytes and paraquat-treated plants, respectively. The applications of these non-GC assays to cryopreserved tissues have yet to be explored.

8.2 Colorimetric Detection of Malondialdehyde (MDA) (Heath and Packer 1968)

One of the consequences of lipid peroxidation is that many breakdown products are produced. In addition to volatile hydrocarbons, malondialdehyde (MDA) is one of the most readily detectable products of lipid peroxidation. MDA has been frequently used as an indicator of freezing/chilling damage in the frozen food industry and in mammalian low-temperature biology. MDA forms a coloured complex with thiobarbituric acid (TBA), which can be detected by either spectroscopy or fluorimetry. As TBA may react with other lipid peroxidation products, it is more appropriate to express the assay on the basis of general TBA reactivity. The spectroscopic TBA assay has been applied to rice tissue cultures and cryopreserved cell suspensions (Benson et al. 1992 a,b). Jouve et al. (1993) used the spectroscopic assay for MDA to assess chilling sensitivity in coffee. Their results show that MDA is a reliable indicator of cold sensitivity coffee genotypes which show differential tolerance to cold treatments. The fluorimetric TBA assay (see method 3) has also proved suitable for plant tissue cultures (Benson et al. 1992b; Benson and Roubelakis-Angelakis 1994). The advantage of this approach is that the TBA complexes are partitioned into a solvent phase, reducing the possible interference from water-soluble complexes.

Objective

- To use MDA formation as a means of assessing freezing injury.

Equipment and materials

- spectrophotometer
- boiling water bath
- ice
- bench top centrifuge

Stock solution

TBA reagent: 0.5%(w/v) thiobarbituric acid in 20%(w/v) trichloroacetic acid made up freshly and stored in the dark

Procedure

1. Weigh out 0.1g tissue in a centrifuge tube.
2. Add 1.4ml of distilled water and vortex.
3. Add 1.5ml of TBA reagent.
4. Boil tubes in a water bath for 25min.
5. Remove and stop the reaction by putting the tubes on ice for 5min.
6. Centrifuge at 2500rpm (1000g) for 10min to remove the cell debris.
7. Measure the absorbance of the supernatant at 532 and 600nm against a reagent blank. Smaller quantities of tissue can be used (e.g. 0.05g in a volume of 1.45ml distilled water).

Results

Subtract the non-specific absorbance at 600nm from the absorbance value at 532nm. Calculate the MDA content using the extinction coefficient at 532nm (155mM/cm) and express results as μM MDA.g^{-1} fresh weight.

Important notes

TBA may form coloured complexes with sugars, polyols and aldehydes. As these compounds are used in tissue culture media and cryoprotectant solutions, it is important to determine if they interfere with the detection of MDA. Absorbance scans performed on common tissue culture media and cryoprotectants indicate that the absorbance of these complexes is not within the 600–532nm absorbance range used to detect MDA (E.E.

Benson and P.T. Lynch, unpubl. observ.). However, it is advisable to routinely check for background interference in media controls as well as using reagent blanks to zero the instruments. Kosugi and Kikugawa (1989) have shown from lipid studies that the coloured TBA complexes absorbing in the 450–550 nm range may be reactants of additional peroxidation breakdown products. These workers suggest that, although the TBA assay is not entirely specific to MDA, it is a very good assay for determining general lipid peroxidation phenomena. For this reason, investigators are increasingly expressing the results of the TBA assay as MDA equivalents (using the extinction coefficient for MDA) or as total TBARS (Thiobarbituric Acid Reactive Substances) on a g fresh weight basis. Alternatively, pure MDA can be determined using high-performance liquid chromatography techniques (Draper and Hadley 1990). Similarly, the TBA-MDA chromophore can be detected directly using HPLC (Li and Chow 1994).

8.3 Fluorimetric Detection of Malondialdehyde (MDA) (Fraga et al. 1988)

MDA may also be detected using a fluorimetric assay. Fraga et al. (1988) have used this technique in mammalian tissues to determine the MDA content of the media in which the tissues are suspended. This assay may also be used to determine the MDA content of tissues directly.

Objective

- To use MDA formation as a means of assessing freezing injury.

Equipment and materials

- spectrofluorimeter
- boiling water bath
- ice
- bench top centrifuge

Stock solutions

- 3% (w/v) sodium dodecyl sulphate (SDS)
- 1N HCl
- 10% (w/v) phosphotungstic acid (store at 4 °C)
- 0.7% (w/v) thiobarbituric acid (may require slight warming to form a solution, make up fresh and store in the dark)
- Butan-1-ol

Procedure

1. Place 0.05–0.1g tissue samples into centrifuge tubes.
2. Add 0.5 ml SDS and vortex.
3. Add 2 ml 0.1 N HCl.
4. Add 0.3 ml of phosphotungstic acid.
5. Add 1.0 ml TBA reagent.
6. Vortex.
7. Heat mixture for 30 min in a boiling water bath.
8. Remove and cool on ice.
9. Extract the TBA/MDA complex into 5 ml of butan-1-ol; this may be achieved by gently inverting the tubes until the coloured pink complex has been transferred to the solvent layer.
10. Centrifuge at 1500 rpm (400 g) for 10 min to remove cell debris and aid the separation of the organic and aqueous phases.
11. Remove the top organic layer and place into a clean tube.
12. Measure the fluorescence emission of the TBA/MDA complex at 555 nm using an excitation wavelength of 515 nm. Butan-1-ol is used as a solvent blank; within experiments, retain a constant scale (e.g. ×1) to standardise fluorescence emission.

Results

MDA content is expressed as fluorescence emission at 555 nm per gram fresh weight. Alternatively, a standard curve can be constructed for malondialdehyde.

Important notes

See Section 8.2

8.4 Fluorimetric Determination of Schiff's Bases (Lunec and Dormandy 1979)

MDA and other aldehyde breakdown products of lipid peroxides are highly toxic to the cell as they can cross-link with proteins and DNA to form Schiff's bases. In badly damaged cells (i.e. during post-thaw recovery), the accumulation of these "secondary" products of free radical damage may interfere with recovery processes. In the long term this may have an affect on genetic stability as secondary lipid peroxidation products have been shown to exhibit both mutagenic and carcinogenic properties (Frankel

1987). Schiff's bases have been routinely used as an index of cold-induced oxidative stress in ischaemic organs, and have been important in evaluating the protective role of various additives (Fuller and Green 1986; Green et al. 1986; Gower et al. 1987). Fluorescent compounds with spectral characteristics of "age-related" Schiff's bases have been detected in plant tissue cultures systems (Benson and Roubelakis-Angelakis 1992, 1994; Benson et al. 1992b).

Objective

- To use Schiff's bases as an indicator of freezing damage, particularly during long-term recovery and re-growth.

Equipment and materials

- spectrofluorimeter
- microfuge
- bench top centrifuge
- ice

Stock solutions

- Tissue extraction buffer (adapted from Murphy and Huerta 1990):
  - 1.0mM $CaCl_2$
  - 0.1mM KCl
  - 0.05M Potassium phosphate buffer (pH 7.0)
  - 0.001M EDTA

Make the phosphate buffer to pH 7.0 by adding KH_2PO_4 to K_2HPO_4 and add the other components to the required concentration in a known volume of buffer (e.g. 100ml), use buffer chilled to 4°C on ice.

- Lipid extraction medium: 2:1 chloroform:methanol (heated to 45°C)

Procedure

1. Place 0.05–0.2g of tissue into an Eppendorf tube.
2. Add 1ml of tissue chilled extraction buffer and rapidly macerate the tissue on ice with the aid of a glass rod.
3. Centrifuge in a microfuge at 4°C for 10min at 10,000rpm (10,000g).
4. Remove the supernatant.
5. Place the pellet in a centrifuge tube containing 6.0ml of chloroform/methanol previously heated to 45°C.
6. Vortex the mixture for 2min.

7. Centrifuge at 1000 rpm (100 g) for 5 min to remove the protein precipitate.
8. Transfer the supernatant to a clean tube.
9. Add 2 ml of deionised water and invert the tube slowly to separate the aqueous and solvent fractions.
10. Centrifuge for 3 min at 1000 rpm (100 g) to aid the separation.
11. Remove the top "aqueous" layer and place in a clean tube.
12. Remove any precipitated debris that has accumulated at the interface.
13. Remove the lower solvent layer and place in a clean tube.
14. Add 50 μl of pure methanol to the solvent layer to clear the emulsion.
15. Immediately measure the fluorescence emission in both the aqueous and solvent fractions in the 400–500 nm region, using an excitation wavelength of 360 nm.
16. Alternatively, the extracts can be stored at −20 °C in the presence of butylated hydroxytoluene (to a final concentration of 2.5 mM).

Results

Schiff's base content may be expressed as fluorescence emission per g fresh weight, if a fixed scale (e.g. ×1) is used for all samples. Alternatively, fluorescence can be expressed in terms of a known fluorescence standard such as quinine sulphate (10 mM in 100 mM H_2SO_4).

Recommendatio

The determination of Schiff's base activity in plant tissues has received little attention, even though it has been a useful indicator of low-temperature-induced stress in mammalian tissues. It is therefore advisable to perform fluorescence scans on plant tissue extracts to determine the major emission maxima in the 400–500 nm region before commencing routine analysis. Preliminary findings with frozen rice cells have confirmed the presence of fluorescence products in chloroform/methanol and aqueous extracts using the methods described above (E.E. Benson and P.T. Lynch, unpubl. observ.). However, there is a possibility that plant tissues may contain phenolic compounds which could interfere with the detection of fluorescent lipid peroxidation products. Whilst Schiff's base components attributed to oxidative stress have been detected in aqueous extracts from animal tissues (Lunec and Dormandy 1979), there exists the possibility, in plant tissues, of co-extracting interfering phenolics in this phase. Thus, analysis of solvent phase extracts (after several aqueous phase washes) may prove a more definitive evaluation of fluorescent lipid peroxidation in plant tissues. Meir et al. (1992) recommend the use of silica Sep-Pak columns to remove potentially interfering compounds from plant extracts.

8.5 Determination of Sulphydryl Group (SH) Status (Chevrier et al. 1988)

One of the earliest proposed mechanisms of freezing injury was termed the sulphydryl-disulphide theory of freezing damage (Levitt 1962). Thus, the freeze-induced destabilisation and oxidation of SH groups in macromolecules causes denaturation and conformational changes which impair function. It is clearly evident that the mechanisms of freezing injury are highly complex and involve many different molecular mechanisms. However, SH groups are indeed vulnerable to freezing, and assays to determine SH content during freeze/thaw cycles and recovery may be useful. The following assay has proved a rapid and simple method when applied to plant tissue culture systems (Benson and Roubelakis-Angelakis 1994), and is based on the methodology of Chevrier et al. (1988).

Objective

- To determine the effects of freezing on SH content.

Equipment and materials

- spectrophotometer
- high speed centrifuge
- microfuge
- ice

Stock solutions

- 1.0M Tris-HCl pH 8.2 (chilled to 4 °C on ice)
- 0.2M Tris-HCl pH 8.5 (chilled to 4 °C on ice)
- 5% (w/v) Trichloroacetic acid (TCA)
- 5% (w/v) sodium dodecyl sulphate (SDS)
- 10mM 5,5′-dithiobis-(2-nitrobenzoic acid) (DTNB) in methanol
- ice cold 100% methanol
- μg range of GSH for calibration curve

Procedure

8.5.1 Total SH

Perform all extractions on ice.

1. Place 0.05–0.1g tissue into a centrifuge tube.
2. Add 2ml 1.0M Tris-HCl pH 8.2.

3. Add 0.5 ml SDS.
4. Add 0.1 ml DTNB (in methanol).
5. Vortex for 5 min to solubilise tissue.

5a. Extraction *modification* for "woody plants" fibrous explants etc. Complete extraction of the tissue is not achieved by using the above protocol, which is suitable for cell suspensions and protoplasts. In the case of less readily extractable tissues, it is necessary to rapidly macerate or homogenise the sample under chilled conditions at stage 3, then add the DTNB and proceed to stage 5.

6. Incubate at room temperature for 20 min with intermediate vortexing.
7. Precipitate the DNA and proteins by adding 3 ml of ice cold methanol.
8. Centrifuge at 10,000 rpm (10,000 g) for 10 min.
9. Remove the yellow supernatant and place in a clean tube; in the case of callus tissue it may be necessary to centrifuge for longer to clear the supernatant.
10. Measure the absorbance at 412 nm against a reagent blank.

GSH Standard Curve

To ensure accurate determinations of SH standard calibrations should be made for each assay method using the same procedures as for the test samples. Thus, for assay 1 the calibration is constructed using 5–50 μg amounts of GSH, added in 5–50 μl aliquots to a total volume of 2 ml 1.0 M Tris-HCl buffer. The assay is then followed as described in steps 5–10 with the omission of the centrifugation steps, but taking into account the additional development time.

8.5.2 Non-Protein SH

1. Place 0.1 g tissue in a centrifuge tube.
2. Add 1.5 ml chilled 5% (w/v) TCA.
3. Vortex.

3a. As with the previous assay (sect 8.5.1), a *modification* of this extraction is suitable for cells and protoplasts. For less readily extractable tissues, it may be necessary to macerate rapidly the tissue in chilled TCA to permit total extraction.

4. Incubate the extract on ice for 10 min.
5. Centrifuge at 10,000 rpm (10,000 g) for 10 min.
6. Transfer 1.0 ml of the TCA supernatant to a clean tube.
7. To this add 3 ml of 0.2 M Tris. HCl pH 8.5.
8. Add 0.1 ml DTNB reagent.
9. Incubate at room temperature for 20 min.
10. Measure the absorbance at 412 nm against a reagent blank.

GSH Standard Curve
SH content is calibrated for 5–50 μl aliquots of GSH (equivalent to 5–50 μg GSH) in a total volume of 1.0 ml TCA. The assay is performed according to stages 4–10 with the omission of the centrifugation steps, but taking into account the additional development time.

Results SH content is expressed as μg GSH equivalents.g^{-1} fresh weight.

Recommendations When applied to in vitro cultures the SH content of the tissues was found to be highly variable (on a fresh weight basis) dependent on the tissue type (E.E. Benson, pers. observ.). Thus, for organised, cytoplasmic tissue the assay can be performed on small samples (0.05 g). However, in the case of vacuolated, less meristematic tissue (cells and callus) the fresh weight of the tissue must be higher (0.2 g) to be within the limits of assay detection. The protein fraction of the SH tissue content can be determined by subtracting non-protein SH from total SH.

8.6 Determination of Viability Using the Fluorescein Diacetate Vital Stain (Widholme 1972)

The interpretation of assays used to assess cryopreservation injury must be dependent, in part, on the assessment of post-freeze survival and recovery. Thus, viability tests are invaluable in aiding the interpretation of more specific biochemical tests of injury. However, some caution must be taken in interpreting viability tests. The time at which they are performed is critical. A test carried out immediately after thawing will give an indication of the ability of the cells to survive freezing injury. However, a second assay should be done within 24 h in order to avoid false positives. A time course of viability is useful in optimising recovery conditions and will provide corroborative data for assays used to assess damage. When possible, viability tests should be supported by visible signs of re-growth (developmental changes, cell division, fresh weight gain).

Objectives

The FDA viability stain is suitable for cells and organised tissues which are thin enough to be viewed under a UV microscope. It is not suitable for thick and optically opaque tissues such as shoot-tip meristems. It has, however, been successfully used to determine post-freeze viability in hairy root cultures (Benson and Hamill 1991). Viable cells containing esterase activity cleave the stain which then fluoresce intensely yellow/green under a UV microscope. Non-viable cells do not fluoresce.

Equipment

- UV microscope

Stock solution

- Fluorescein diacetate [FDA, 0.1% (w/v)] in 100% acetone store in a refrigerator at 4 °C. Add a few drops of the stock stain to 10 ml of routine liquid culture medium just before use.

Stain procedure

1. Put a small amount of cells or root tips onto a clean microscope slide.
2. Add a few drops of diluted stain and leave for 2–3 min.
3. Count the number of cells in a field of view under a UV microscope.
4. Count the number of cells under the same field of view under bright fields.
5. Count the viability in at least five fields.

Results

Viability is expressed as % fluorescence by comparing the number of fluorescent cells to the total bright field cell population.

8.7 Triphenyl Tetrazolium Chloride (TTC) Spectroscopic Test for Viability (Steponkus and Lamphear 1967)

This method is suitable for most tissue types. However, it may be especially useful for samples which are difficult to assess using FDA staining. If adequate infiltration of TTC is achieved under vacuum, it may be applied to organised structures. The TTC dye is reduced by viable cells and an insoluble red formazan product is formed which can be measured spectroscopically.

Equipment

- spectrophotometer
- boiling water bath
- bench top centrifuge

Stock solutions

- TTC [0.6% (w/v)] and 0.05% Tween 80 in: 0.05 M Na_2HPO_4/KH_2PO_4 buffer pH 7.4
- 95% (v/v) ethanol

Procedure

1. Place 0.05–0.1 g of tissue into a centrifuge tube.
2. Add 3 ml of TTC reagent.
3. Infiltrate under vacuum for 10 min or longer for dense tissue.
4. Incubate overnight at 28 °C.
5. Centrifuge the cells (if necessary) and remove the TTC solution.
6. Wash the cells with distilled water.
7. Resuspend the cells in 7 ml 95% (v/v) ethanol.
8. Extract the formazan by boiling the tubes in a water bath for 5–10 min.
9. Vortex the samples and centrifuge at 2500 rpm (1000 g) for 10 min.
10. Measure the absorbance of the supernatant against a reagent blank.

Results Express the results as absorbance (490 nm) per g fresh weight. Viability is often expressed as a % of an unfrozen control.

B. Molecular Methods

There is a considerable amount of literature available on the use of techniques in molecular biology. The objective of this section is to provide a bench-top working text, where molecular procedures can be readily applied in genetic stability assessments of plants and tissues recovered from cryopreservation. The information and techniques described do not compete with the established molecular text books, but provide a practical guide for newcomers to this subject. The molecular analysis of DNA involves a series of detailed procedures from the isolation of DNA to the analysis of results on an X-ray film. The flow diagram (Fig. 2) is intended as a guide for the less experienced investigator, and shows the different steps and connections between procedures involved in the analysis of DNA. There is a recommended reading list at the end of this chapter for investigators wishing to gain more theoretical and practical details.

The use of specialised laboratory facilities for radioactive work associated with molecular biology is no longer essential. The reduced need for

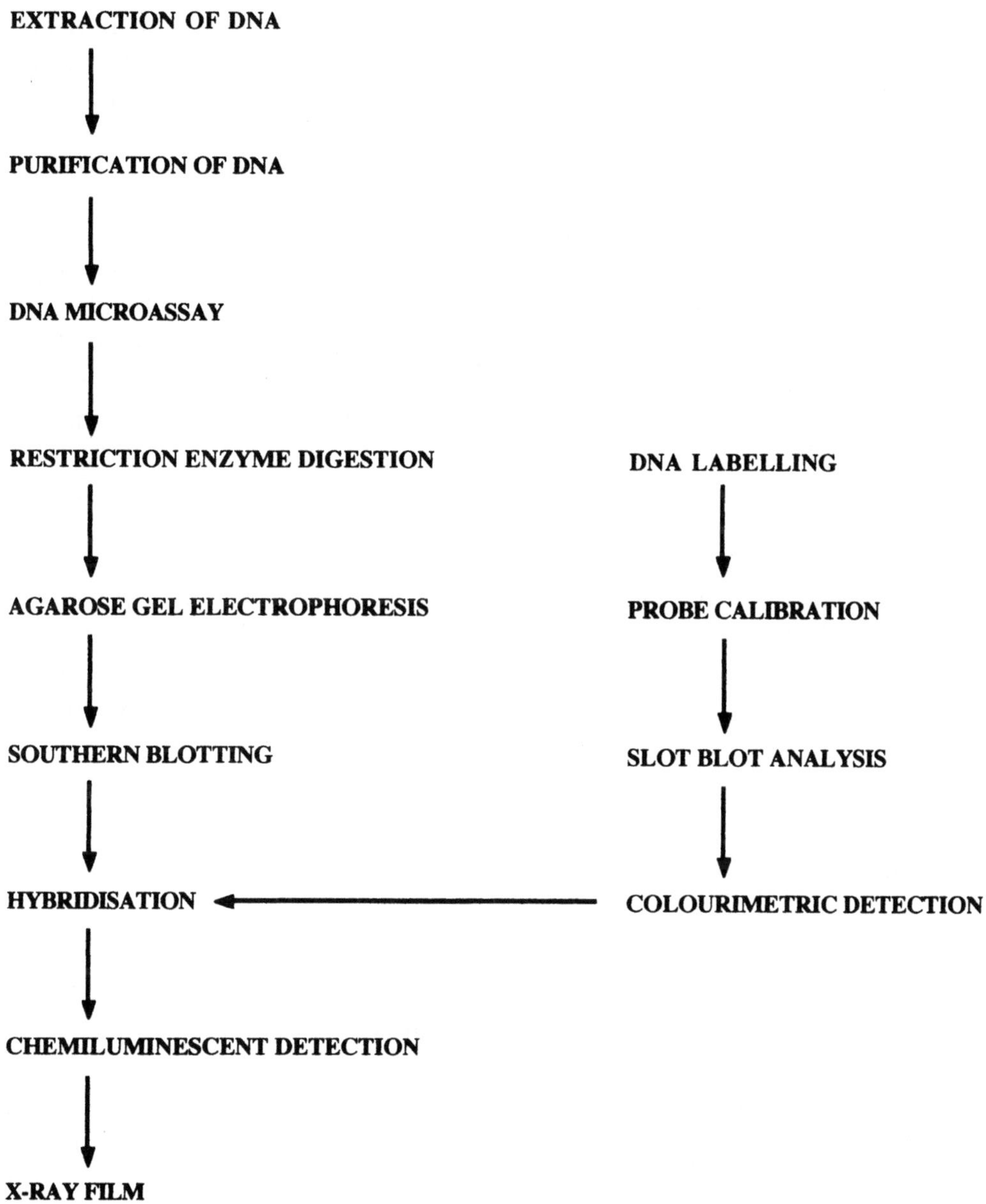

Fig. 2. Flow diagram showing the connections between the different procedures that are used in the molecular analysis of DNA extracted from plants or tissue cultures recovered from cryopreservation

specialised facilities is largely due to the impact of non-radioactive DNA labelling procedures. In this section, the use of biotin labelling, the detection of DNA sequences by colorimetric and/or chemiluminescent procedures are described (Harding 1992). Non-radioactive procedures are making molecular techniques increasingly available to a range of investigators with broad research interests. Furthermore, there will be a need for these techniques with the anticipated development of in vitro gene banks. It is the hope of the authors that the procedures in this section will be especially useful to scientists wishing to pursue molecular studies without the hazardous use of radioactivity.

Investigations into the genetic stability of plants recovered from cryopreservation may be studied in several ways using the tools of molecular biology. The use of ribosomal RNA gene (rDNA) probes described in this section represents an approach that can be applied to a range of plant species. The use of a heterologous probe (derived from wheat) to restriction enzyme map the rDNA of *Solanum tuberosum* (potato) has been described (Harding 1991a). This study was possible due to the highly conserved nature of the rDNA coding regions across the plant kingdom. The degree of rDNA conservation between monocotyledonous and dicotyledonous plants is shown by the 93.5% homology in the 18S gene of soybean and maize (Eckenrode et al. 1985). The use of rDNA probes as molecular tools to investigate the stability of plants regenerated from medium (slow growth conditions) and long-term (cryopreservation) storage has been reported (Harding 1991b, 1994). Techniques which underpin these stability studies are described below.

8.8 Mini DNA Extraction Procedure (CTAB Method)

The isolation of plant DNA is possible using a variety of methods. The procedure described is a modified method described by Rogers and Bendich (1988) and is applicable to a wide range of species.

Equipment and materials

- bench-top microfuge
- eppendorf tubes 1.8 ml
- glassware for reagents (Schott bottles)
- liquid nitrogen Dewar
- mortar and pestle

Stock solutions

Solutions marked with an * can be autoclaved (121 °C for 20 min) to prolong their self-life:

- EDTA* 0.5 M (MW 372.24): Dissolve 186.12 g in 400 ml of distilled water with 10 g NaOH pellets. Adjust to pH 8.0 and make to 500 ml with distilled water.
- NaCl* 5 M (MW 58.44): Dissolve 145 g in 400 ml of distilled water and adjust to a final volume of 500 ml.
- Tris buffer* 1 M (MW 121.1). Dissolve 60.55 g in 400 ml of distilled water. Adjust the pH by adding an approximate volume of concentrated HCl. If the pH exceeds the desired value, re-adjust with 40% NaOH.

pH required	HCl added (ml)
7.0	50
7.5	32
8.0	20

Make up the volume to 500 ml with the use of a volumetric flask.

Reagents and solutions

- Extraction buffer (100 mM Tris HCl pH 8.0; 20 mM EDTA; 1.4 M NaCl; 2% CTAB and 1% PVP (MW 40,000)
- chloroform and isoamyl alcohol (24:1)
- 10% CTAB-salt solution (10% CTAB in 0.7 M NaCl)
- precipitation solution (50 mM Tris HCl pH 8.0; 10 mM EDTA and 1% CTAB)
- high salt buffer (10 mM Tris HCl pH 8.0; 5 mM EDTA and 1 M NaCl)
- isopropanol (propan-2-ol)
- 70% ethanol
- TE buffer (10 mM Tris HCl pH 8.0; 1 mM EDTA)
- RNase A (1 mg ml^{-1})

Procedure

1. Select green healthy leaf material (0.5 g).
2. Grind leaves in liquid nitrogen using a mortar and pestle.
3. Transfer powder into an Eppendorf tube (1.8 ml) and add 0.5 ml extraction buffer pre-heated to 65 °C.
4. Add an equal volume of chloroform/isoamyl alcohol (24:1) and emulsify.
5. Centrifuge at 10,000 rpm (10,000 g) in a microfuge for 5 min.

6. Remove the aqueous phase and add 1/10 volume of 10% CTAB-salt solution.
7. Extract with an equal volume of chloroform/isoamyl alcohol and centrifuge at 10,000 rpm (10,000 g) for 5 min.
8. Remove the aqueous phase, add an equal volume of CTAB precipitation solution and mix.
9. Nucleic acid may be visible as a fibrous material, centrifuge at low speed at 5000 rpm (2000 g) for 5 min.
10. Resuspend pellet in 0.5 ml high salt buffer.
11. Remove particulate material by centrifugation at 10,000 rpm (10,000 g) for 5 min and transfer the DNA solution into a fresh Eppendorf tube (1.8 ml).
12. Add 0.8 volume of cold isopropanol and centrifuge the pellet at low speed, 5000 rpm (2000 g) for 5 min.
13. Wash the pellet in 0.5 ml of cold 70% ethanol.
14. Allow the pellet to air dry (invert the tube) and resuspend pellet in 100–200 μl TE buffer.
15. To remove RNA, the preparation may be treated with 1 μl RNase A for 30 min at room temperature.
16. The DNA preparation should be sufficiently pure for digestion by restriction enzymes. If the DNA requires further purification use either the resin column method (Qiagen) or caesium chloride/ethidium bromide procedure.

Important notes Advice in these notes relating to a specific point has been numbered to correspond to the same stage in the experimental procedure. This procedure may be scaled up for the isolation of large amounts of DNA by increasing proportionately the reagent volumes.

1. Soiled leaves may be briefly washed in tap water, 20 mM potassium chloride, 5 mM EDTA and finally, distilled water.
2. Ensure that the material is ground to a fine powder in the liquid nitrogen.
3. The extracted mixture should be viscous.
4. Emulsify the mixture by gentle shaking.

5. After centrifugation there should be two distinct phases, the top layer is the aqueous phase.
6. Ensure the 10% CTAB-salt is completely mixed with the aqueous phase.
7. Emulsify the mixture by gentle shaking.
8. Nucleic acids should be precipitated at this stage; however, for some species a precipitate is not always observed; this may be facilitated by cooling the sample to 15 °C (not below, as CTAB will precipitate) for 15 min.
9. Avoid compacting the nucleic acids into a tight pellet.
10. The time for the pellet to dissolve is variable, and depends on the yield of DNA and the degree of impurities. Heating the sample to 65 °C will aid dissolution.
11. Many impurities do not dissolve in the high salt buffer, and are removed at this stage.
12. The cold isopropanol (propan-2-ol) treatment should precipitate white fibrous strands of DNA.
13. Washing the pellet in cold 70% ethanol further removes contaminating material (CTAB and salt).
14. Remove all traces of ethanol before dissolving the pellet. The time for dissolution depends on the quality of the DNA, high molecular weight DNA takes longer to dissolve.
15. This extraction procedure precipitates total nucleic acids, RNA is removed by treatment with RNase A (previously heated to 80 °C for 15 min to remove DNase activity).
16. This extraction procedure produces high molecular weight DNA from a range of plant species, that can readily be digested by several restriction enzymes. In the case of woody species, and species containing high levels of phenolics, tannins, gums and latex, additional purification methods possibly need to be employed.

8.9 Purification of DNA Extracts

The extraction of DNA from many species can be achieved by the mini DNA extraction procedure described. However, the isolation of DNA from plants which produce high levels of natural (secondary) products, especially woody species, is problematical, and DNA preparations often require further purification. This can be achieved by one (or a combination) of the described methods.

Resin column procedure uses the Qiagen columns (containing a modified ion exchange resin) which are commercially available as tubes or tips (which conveniently fit most Gilson or equivalent pipetmans) and the procedure described has been modified for the purification of plant DNA.

DNA purification by the caesium chloride-ethidium bromide centrifugation procedure can be achieved with a variety of swing-out, fixed-angle or vertical rotors and ultracentrifuges. DNA is centrifuged in a self-generating caesium chloride gradient in the presence of saturating levels of ethidium bromide (which intercalates with the DNA). The complex separates in the gradient at an equivalence position between the density of the caesium chloride and the buoyant density of the DNA-ethidium bromide complex. This DNA purification procedure is especially suitable for a scaled up version of the mini DNA extraction procedure, where large amounts of DNA require further purification.

Equipment and materials

- Abbe' refractometer
- analytical balance
- glass/plastic pipettes or Gilson pipetmans
- long wavelength UV lamp
- magnetic stirrer
- polycarbonate tubes
- syringes 2/5 ml and G19 gauge needles
- ultracentrifuge, rotor and centrifuge tubes

Stock solutions

- 0.5M EDTA
- 1M Mops (MW 209.26). In duplicate dissolve 21 g in 80 ml of distilled water, adjust the solutions to pH 7.0 and 8.0 and make to 100 ml with distilled water
- 5M NaCl

Reagents and solutions

- buffer 1 (750mM NaCl, 50mM Mops, pH 7.0 and 15% ethanol)
- buffer 2 (1000 mM NaCl, 50 mM Mops, pH 7.0 and 15% ethanol)

- buffer 3 (1500 mM NaCl, 50 mM Mops, pH 8.0 and 15% ethanol)
- ethanol
- TE buffer (10 mM Tris HCl pH 8.0; 1 mM EDTA)
- caesium chloride
- ethidium bromide 10 mg ml^{-1}
- isopropanol (propan-2-ol)

8.9.1 Resin Column Procedure (Qiagen)

Important notes

- This procedure is appropriate for the purification of DNA samples prepared form tissues by the mini DNA extraction procedure.
- These columns are normally supplied with a manufactures' hand book, that describes the basic procedures.
- Prior to the purification of valuable plant DNA samples, technical expediency can be obtained by a practice run using lambda DNA.

Important notes

- After the final elution do not discard the column. If DNA is not recovered and precipitated, the DNA is possibly still bound to the column. Increasing the salt concentration and/or the pH will release the DNA from the column.

Procedure

1. Equilibrate the column in 1 ml of buffer 1, by passing the buffer solution through the column three times.
2. Adjust the DNA sample to 0.5 ml containing the same composition as buffer 1.
3. Aspirate the equilibrated column three times with the adjusted DNA sample.
4. Wash the column six times with 0.5 ml of fresh buffer 2.
5. Elute the DNA from the column by washing with two 250 µl volumes of buffer 3. Wash each 250 µl volume down the column three times.
6. Precipitate DNA with 2 volumes of cold ethanol (store at −20 °C) and centrifuge at 10,000 rpm (10,000 g) in a microfuge for 5 min.
7. Resuspend the DNA pellet in 50 µl of TE buffer.
8. The concentration of the purified DNA should be assayed before digestion by restriction enzymes.

8.9.2 Caesium Chloride-Ethidium Bromide Centrifugation

Recommendations

- As DNA is lost during the various processing stages, a relatively large sample of DNA is required prior to centrifugation.
- Dissolving DNA in TE buffer at 4 °C reduces nuclease activity.
- As CsCl dissolves the temperature is reduced, to increase the solubility of CsCl heat the tube to 20 °C.
- An ethidium bromide-protein/carbohydrate complex will precipitate during the incubation.
- The precipitate may be a pellet and a floating complex at the surface of the tube; this can be gently scraped to the side of the tube.
- The refractive index must be adjusted at 20 °C or the position of the DNA in the gradient will be altered, possibly leading to loss of the DNA sample.
- The use of the analytical balance to adjust the density of the CsCl gradient must also be performed at 20 °C. This is especially important for the use of vertical rotors where the path length is relative short.
- Ensure the appropriate centrifuge tubes are correctly filled and sealed.
- The temperature, speed and time are critical factors in the establishment of a DNA gradient. Centrifugation between 14 and 16 h for vertical rotors is important; due to the short path length DNA may be sedimented against the side of the tube.
- The rescue of DNA by the method of side puncture is difficult and requires a little practice before removing important DNA samples.
- Ensure the isopropanol (propan-2-ol) is equilibrated with CsCl and TE buffer or DNA may be lost.
- Repeated extraction is required to remove all traces of ethidium bromide.
- Dialysis tubing requires pre-treatment by boiling in 2% sodium carbonate for 10 min, washing in SDW, then boiling in 5 mM EDTA, washing in SDW, and finally autoclaving in TE buffer.
- To prevent the introduction of nucleases, use sterile instruments.
- The precipitation of DNA should be facilitated at low temperature.
- DNA is stable in storage at −20 °C for long periods of time.

Procedure

1. Pool several mini preparations of DNA.
2. Resuspend the DNA in 4 ml of TE buffer at 4 °C.
3. Transfer the TE-DNA solution to a 15 ml plastic tube containing 3.9 g CsCl and dissolve by inverting the tube.
4. Add 0.6 ml of ethidium bromide, mix and leave at 20 °C for 10 min.
5. Remove the precipitate by centrifugation at 10,000 g for 15 min.
6. Adjust the refractive index with the use of an Abbe' refractometer to the index 1.3870–1.3880 with the addition of CsCl or TE buffer.
7. Alternately, use an analytical balance and weigh 10 μl samples of the solution adjust the density to 15.7–16.1 mg (for vertical rotors use 16.1–16.2 mg).
8. Fill and seal the appropriate centrifuge tube.
9. Balance the tubes to within 50 mg and centrifuge the DNA under the following conditions: temperature 20 °C; speed 40,000 rpm and time 15–16 h.
10. Remove the tubes and visualise the gradient under long wavelength UV light. Pierce the top of the centrifuge tube to allow the entry of air. Rescue the band from the caesium chloride gradient with the use of a G19 needle and 2–5 ml syringe by the method of side puncture (see Fig. 3). Place the rescued band into polycarbonate tubes.
11. Extract the ethidium bromide by adding an equal volume of isopropanol

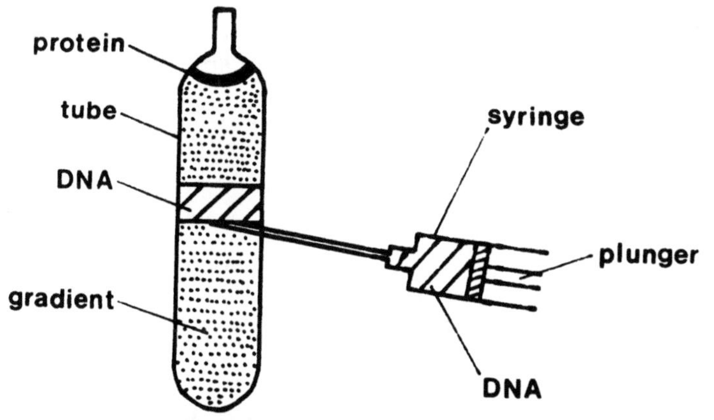

Fig. 3. Diagram showing the recovery of DNA/ethidium bromide complex after buoyant density caesium chloride/ethidium bromide centrifugation in a vertical centrifuge tube by the method of side puncture

(equilibrated with CsCl and TE buffer) and gently inverting the tube. Remove the top layer and discard.

12. Repeat the extraction at least six times.

13. Transfer the DNA into a pre-treated dialysis tube and immerse into 1 litre of cold (4 °C) TE buffer. Place the flask on a stirrer in the cold room. Change the TE buffer after 1 h with pre-cooled TE buffer and leave overnight.

14. Remove the dialysis tubing, blot off the excess buffer, cut the tubing and transfer the contents into an Eppendorf tube. Store the DNA at −20 °C.

15. Alternatively, add an equal volume of TE buffer to the DNA, precipitate the DNA with the addition of 1 volume of isopropanol (propan-2-ol), either store overnight at −20 °C or use a dry-ice-methanol bath and immerse the mixture for less than 5 min.

16. Centrifuge the DNA, allow the DNA pellet to air dry and resuspend in a small volume of TE buffer. Store at −20 °C.

8.10 DNA Microassay

Prior to digestion of the isolated DNA by restriction enzymes, it is important to measure the concentration of DNA for successful Southern blotting. The method described allows the simple, accurate determination of DNA concentration.

Objectives

- There are several assay procedures for the estimation of DNA concentration. The microassay is convenient, cheap and relatively simple.

Equipment

- spectrophotometer
- magnetic stirrer

Reagents and solutions

- TE buffer
- 3M perchloric acid
- diphenylamine reagent (4% diphenylamine, 0.01% paraldehyde dissolved in glacial acetic acid).

Important notes

- The DNA standards can be stored at 4 °C for several months.
- Discoloration of diphenylamine reagent does occur with time, and the reagent should be prepared freshly.

• CAUTION: These reagents are either extremely caustic or toxic – handle with care and use rubber gloves. !

Calibration Curve **Procedure**

1. Prepare a standard solution of calf thymus DNA. Dissolve a known weight of dried DNA (10 mg) in approximately 10 ml of TE buffer (this normally takes 6–12 h with continuous stirring), i.e. × mg DNA y ml^{-1}.
2. Prepare the standard DNA solutions in TE buffer within the range 0, 10, 20, 30, 40, 50, 60, 70, 80, 90 and 100 $\mu g.ml^{-1}$.
3. Add 150 µl of each solution to a separate large Eppendorf tube (in duplicate).
4. Add 150 µl of 3M * perchloric acid.
5. Add 180 µl of * diphenylamine reagent (4% diphenylamine, 0.01% paraldehyde dissolved in glacial acetic acid) and gently mix by inverting the tube.
6. Incubate the tubes in the dark at 30 °C overnight.
7. Determine the optical density at 600 nm and plot the standard curve.

Measurement of Samples

1. The DNA concentration of unknown samples is determined by taking 5, 10 and 15 µl, where appropriate, and making the final volume up to 150 µl with TE buffer.
2. The concentration can be determined by repeating the above assay.

8.11 Restriction Enzyme Digestion

Digestion of DNA with restriction enzymes is normally performed according to the manufacturers' instructions (×10 reaction buffer normally accompanies the restriction enzyme). Restriction enzyme activity is defined as 1 unit of an enzyme will cleave 1 µg of lambda DNA in 1 h at 37 °C. The time of incubation is dependent on the amount of DNA, the purity of the DNA, and the number of restriction enzyme target sequences within the DNA. Complete digestion of plant DNA is normally achieved after 2 to 4 h.

Equipment and materials

- heated water bath
- Eppendorf tubes

Reagents

- Loading buffer (×10 stock contains 0.25% bromophenol blue, 0.25% xylene cyanol and 25% Ficoll (MW 400) dissolved in distilled water).
- 0.5M EDTA
- SDW

Procedure A reaction mixture typically contains the following components:

– DNA	= X μl
– reaction buffer (×10 stock)	= 2 μl
– restriction enzyme (1–10 units)	= 2 μl
– sterile distilled water	= y μl
Total volume	= 20 μl

1. Incubate the reaction mixture at the indicated temperature for 2 to 4 h.
2. Stop the reaction by the addition of 2 μl of 0.5M EDTA.
3. Prior to agarose gel electrophoresis add 1/10 volume (2–3 μl) of loading buffer to the reaction mixture.

Important note

- Digestion of genomic DNA is performed in conditions of enzyme excess. Generally, several units of the restriction enzyme per μg of DNA are used to avoid the generation of partial digestion products, and the possibility of artefactual results.

8.12 Agarose Gel Electrophoresis

Restriction enzyme digested DNA is fractionated into a range of molecular species by electrophoresis in an agarose gel. The agarose concentration is determined by the molecular size of the DNA fragments to be fractionated. Restriction enzymes that recognise 6 bp sequences (e.g. Bam HI, Eco RI and Hind III etc.) cleave genomic DNA into a range of fragments, which are readily fractionated by a 1.2% agarose gel. To assist the analysis of DNA fragments, it is important to include molecular weight markers. This is achieved with the use of lambda DNA (1 μg) digested with Eco RI and Hind III restriction enzymes, that provides a suitable range of markers (21.2 to 0.56 kb). Normally 10 μl of digested lambda DNA (1 μg) diluted one or two orders of magnitude should be loaded in the gel next to the digested DNA samples (see results section). Furthermore, a positive hybridisation control containing digested probe DNA diluted two or three orders of magnitude should be fractionated in parallel with the DNA samples (see important notes).

Equipment and materials

- heated water bath
- horizontal agarose gel electrophoresis apparatus (dimensions 11 × 14 cm)
- microwave or bench steamer
- polaroid camera with polaroid film (ASA 3000) film
- power pack (150 volts, 400 mA)
- UV transilluminator light box
- top pan balance

Stock solution

Tris-borate (TBE), pH 8.0 buffer: The composition of a 1 × solution is 0.089 M Tris, 0.089 M boric acid and 0.002 M EDTA. To prepare a 5 × solution, dissolve 54 g of Tris (MW 121.1), 3.72 g of EDTA (MW 372) and 27.5 g of boric acid (MW 61.83) in 750 ml of distilled water. Adjust pH to 8.0, and final volume to 1 l with distilled water.

Reagents

- ultra-pure agarose (BRL)
- ethidium bromide 10 mg · ml^{-1}

Important notes

- Prior to loading the gel, lambda DNA standards are heated to 65 °C for 10 min to denature certain DNA fragments.
- The use of a positive hybridisation control is important in experiments to enable hybridisation between the probe and filter bound DNA to be determined. This control is required in the use of heterologous probes, where DNA samples may lack hybridisation signals.
- Horizontal agarose gel electrophoresis may be performed with a range of different sized gel kits.
- Caution should be taken with the use of a microwave as super-heated agarose may spout from the flask.

- Prior to electrophoresis ensure that there is 5 to 10 mm of buffer above the surface of the gel. This prevents or reduces fragment distortion caused by heat from the current.
- Caution should be taken when handling ethidium bromide solutions as they are mutagenic.
- Gels may also be photographed with a conventional 35 mm camera in conjunction with a red filter, and black and white film (FP4, Ilford).

Procedure

1. A medium-sized horizontal agarose gel electrophoresis apparatus (dimensions 11 × 15 cm approximately) may be filled with 100 ml of agarose.
2. To cast an agarose gel, first seal the open edges of the gel base with masking tape and fit the comb.
3. Prepare 1.2 g of agarose with 100 ml of TBE buffer in a conical flask, and note the total weight.
4. Melt the agarose with occasional stirring in a microwave or bench steamer.
5. While the agarose is molten, add distilled water to the original weight, and place the flask in a water bath at 55 °C.
6. Place the gel base onto a horizontal platform, and pour the cooled agarose into the gel base.
7. After the gel has set, remove the comb and masking tape, and place the base into the electrophoresis tank.
8. Gel electrophoresis is performed in a 1 × solution of Tris-borate (TBE) pH 8.0 buffer at 3–5 $V cm^{-1}$ for 12–15 h.
9. After electrophoresis DNA in the gel is visualised by staining with ethidium bromide (1 $\mu g.ml^{-1}$ final concentration) solution for 20–30 min, and illuminating the stained gel with UV light from a transilluminator light box (see Fig. 4).
10. The gel may be photographed with a Polaroid camera and red filter (ASA 3000 film).

8.13 Southern Blotting

Prior to hybridisation, it is necessary to transfer the fractionated DNA from the gel to a membrane filter. This transfer requires double stranded

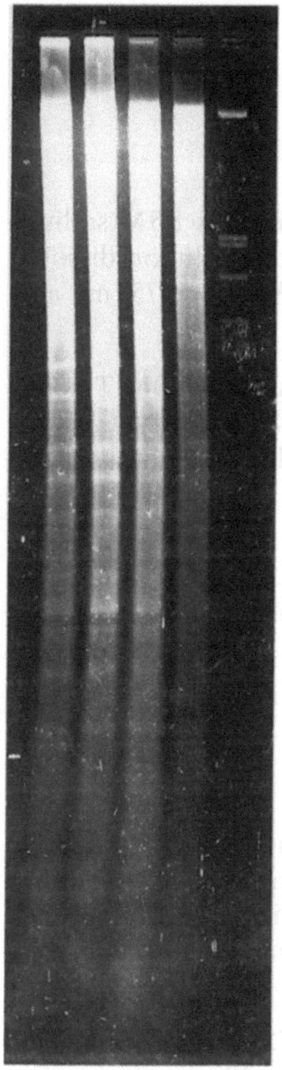

Fig. 4. An agarose gel containing digested DNA stained with ethidium bromide visualised under UV light. The DNA tracks to the left hand side are duplicate samples of *S. tuberosum* cultivar Desiree, those to the right are Golden Wonder digested with Eco RI. Lambda DNA digested with Eco RI and Hind III restriction enzymes is shown to the far right

DNA fragments to be converted to single-stranded fragments, which bind to most hybridisation membranes making the DNA accessible for hybridisation.

Equipment and materials

- bench top tilt shaker
- 3MM chromatography paper (Whatman)
- heated vacuum oven
- paper towels

- perspex or glass plates (20 × 25 cm)
- plastic dishes (20 × 30 × 5 cm)
- nylon membrane filters (Tropilon, Tropix)

Stock solutions

- *Denaturation Solution.* The composition of this solution is 1.5 M sodium chloride and 0.5 M sodium hydroxide. To prepare this solution dissolve 87.66 g of NaCl (MW 58.66) and 20 g of NaOH (MW 40) in 750 ml, and adjust the volume to 1 l with distilled water.
- *Neutralisation Solution.* The composition of this solution is 1.0 M Tris pH 7.5 and 1.5 M NaCl. To prepare this solution dissolve 87.66 g of NaCl (MW 58.66) and 121.1 g of Tris (MW 40) in 750 ml of distilled water. Adjust pH to 7.5, and final volume to 1 l with distilled water.
- *20 × SSC Solution.* The composition of 1 × SSC is 0.15 M NaCl, 0.015 M tri sodium citrate, pH 7.0. To prepare a 20 × solution dissolve 175.3 g of NaCl (MW 58) and 88.2 g of sodium citrate (MW 294.1) in 750 ml of distilled water. Adjust pH to 7.0, and final volume to 1 litre with distilled water.
- 0.25M HCl

Procedure

1. The transfer of high molecular weight DNA fragments is facilitated by depurination of the DNA. Incubate the gel in 250 ml of 0.25 M HCl for 10 min using a tilting platform shaker. Larger or thicker gels require an increase in the incubation time for depurination, denaturation and neutralisation.
2. Remove the HCl, the conversion to single stranded DNA is performed by placing the gel in 250 ml of denaturation solution for 30 min.
3. Remove the denaturation solution, and place the gel in 250 ml of neutralisation solution for 30 min prior to Southern blotting.
4. Gels are capillary (Southern) blotted (Southern 1975) in the presence of 20 × SSC onto a suitable membrane. It is important at all stages in the construction of a Southern blot that air bubbles are removed. Especially those between the gel and the membrane. Care should be taken to ensure that the filter is aligned with the origin of the gel, as this position affects the estimations of molecular weight. It is important to use neutral nylon membrane filters (Tropilon, Tropix) for the chemiluminescent detection of DNA.

!

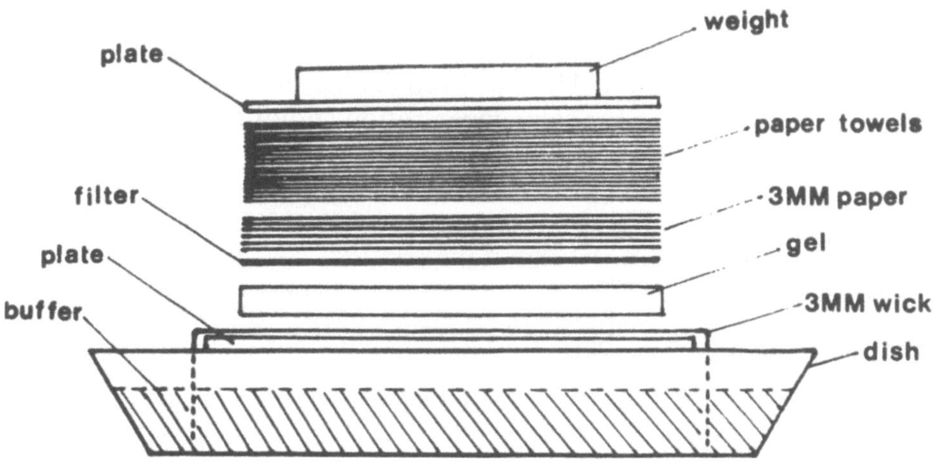

Fig. 5. Diagram showing the construction of a Southern blot (see text for details)

5. Remove the excess agarose, which does not contain DNA, including the agarose above the samples wells created by the comb. The position at the top of the gel is designated the origin of DNA migration.
6. Capillary blotting is performed by several sheets of paper (3 MM chromatography paper) acting as wicks, which soak the 20 × SSC from a reservoir (see Fig. 5). The neutralised gel is placed onto the wicks, and the membrane filter placed on top of the gel so the edge of the membrane is aligned with the origin. Five to ten sheets of 3 MM paper (cut to the size of the gel) and 4 to 5 cm of paper towels are carefully placed on top. A sheet of glass is placed onto the stack with a 500–900 g weight and left overnight.
7. Remove the membrane and wash in 2 × SSC and bake in vacu at 80 °C for 2 h to irreversibly bind the DNA to the membrane.

8.14 Hybridisation Conditions

Hybridisation between the probe and filter bound DNA is dependent on:

- the degree of DNA homology
- the conditions of stringency.

The stringency of hybridisation and subsequent washes is determined by the temperature of incubation and the ionic strength of the solution (salt concentration).

Procedure

The hybridisation conditions described in the text enable the heterologous probe (derived from wheat) to bind to the filter bound DNA (potato). The conditions of stringency, 1M NaCl (equivalent to 6 × SSC) and 65 °C are important, as lowering the salt concentration and/or increasing the temperature may reduce the hybridisation signals.

Equipment and materials

- plastic hybridisation bags (Hybaid)
- plastic 5 ml pipettes
- reciprocal shaking water bath

Stock solutions

- 0.5 M EDTA
- Heparin 10 mg.ml^{-1}
- 20 × SSC
- sodium dodecyl sulphate (SDS) 10% solution. Dissolve 10 g in 80 ml of distilled water and heat to 50 °C to help dissolution. Adjust to a final volume of 100 ml
- 5M NaCl
- 1M Tris pH 7.5 buffer

Reagents and solutions

- pre- and hybridisation solution (0.5% PVP-360, 2.0% SDS, 0.2% heparin, 1 mM EDTA, 1 M NaCl and 50 mM Tris HCl pH 7.5)
- Wash solution (0.5% PVP-360, 2.0% SDS, 1 M NaCl and 50 mM Tris HCl pH 7.5)
- 4 × SSC and 0.1% SDS
- 2 × SSC

1. The pre-hybridisation stage is required to block non-specific binding of the probe. Pre-hybridisation of the nylon membranes is performed at 65 °C for 2–4 h in plastic bags containing pre-hybridisation solution in a reciprocal shaking water bath. While sealing the plastic bags during both pre- and hybridisation stages ensure that all air bubbles are removed.
2. After incubation, replace with fresh pre-hybridisation solution.
3. Heat denaturing the biotin labelled probes converts the probe to single stranded DNA. Heat denature (5 min at 100 °C) the biotin labelled probe and inject this into the hybridisation bag. Mix the contents.
4. Hybridisation is performed for 16–20 h under the same conditions as pre-hybridisation.

5. After hybridisation, the filter is washed twice in wash solution at 65 °C for 5 min.

6. Washing the filter after hybridisation at the required temperature is important as this reduces non-specific binding. Wash the filter twice in 4 × SSC, 0.1% SDS for 10 min at 65 °C.

7. Wash the filter once in 2 × SSC at 20 °C for 2 min and semi-dry the filter.

8.15 Biotin DNA Labelling Procedures

Plasmid pTa 71 containing the ribosomal genes derived from wheat (Gerlach and Bedbrook 1979) or isolated gel fragments are biotin labelled by the following methods. Plasmid DNA is labelled by the method of nick translation (NT) essentially described by Rigby et al. (1977) and Sambrook et al. (1989) with the following modifications.

The random priming procedure is a modified method described by Feinberg and Vogelstein (1983, 1984), where biotin-7-dATP is substituted for alpha-^{32}P-dCTP.

Equipment and materials

- Eppendorf tubes
- glass wool
- heated waterbath
- plastic 5 ml pipettes.

Stock solutions

- 0.4 mM Biotin-7-dATP (BRL)
- 1mM dCTP
- 1mM dGTP
- 1mM dTTP
- 1M Tris pH 7.5 buffer
- 1M Tris pH 7.0 buffer
- 1M Tris pH 8.0 buffer
- 0.5 M EDTA
- 1M Magnesium chloride* (MW 203.31). Dissolve 101.65 g in 400 ml of distilled water and adjust to a final volume of 500 ml
- 5M NaCl

- 10% SDS

Reagents and solutions

- NT (×10) buffer (0.5 M Tris.HCl pH 7.2, 0.1 M $MgSO_4$, 1 mM dithiothreitol, 500 μg.ml^{-1} BSA)
- 0.15 M NaCl in 50% glycerol
- 10 mM Tris.HCl pH 7.5 buffer
- 1% dextran blue solution
- 10 mg.ml^{-1} BSA solution
- enzymes: DNase I, polymerase I and Klenow fragment
- pancreatic DNase I (Sigma) dissolve 1 mg (2000 Kuntz units) in 0.15 M NaCl and 50% glycerol to a final concentration of 0.1 mg ml^{-1}, and store at −20 °C
- solution 1 (1.25 M Tris.HCl, pH 8.0 and 0.125 M magnesium chloride)
- solution 2 (1 ml of solution 1 with 18 μl of beta-mercaptoethanol, and 5 μl of 0.1 M dCTP, dGTP and dTTP in 3 mM Tris pH 7.0 and 0.2 mM EDTA)
- solution 3 (2 M Hepes buffer pH 6.6)
- solution 4 (random hexadeoxyribonucleotides in 3 mM Tris pH 7.0 and 0.2 mM EDTA at an optical density of 90–100 units.ml^{-1}).
- stop solution (20 mM Tris.HCl pH 7.5, 2 mM EDTA, 20 mM NaCl, 0.25% SDS, and 1 μM dTTP)
- Sephadex G50 equilibrated in TE buffer

Procedure

8.15.1 Method of Nick Translation

1. Mix DNA (1 μg) with 2.0 μl of ×10 NT buffer.
2. Add 2 μl of biotin-7-dATP, dCTP, dGTP and dTTP to the reaction mixture.
3. Make a 1/1000 dilution of DNase I in 10 mM Tris HCl pH 7.5 buffer and add 4 μl to the reaction mixture.
4. Add 1 μl (5 units) of polymerase I to the reaction mixture.
5. Make the reaction mixture to a total volume of 20 μl.
6. Incubate the reaction mixture at 15 °C for 1 h. Ensure the nick translation reaction is maintained at 15 °C during the 1 h incubation to avoid excessive nicking by the DNase.

7. Stop the reaction by the addition of 2 μl of 0.5 M EDTA, and store at 4 °C.
8. Separation of unincorporated biotin-7-dATP from the biotin labelled DNA is achieved by adding 25 μl of a 1% dextran blue solution to the reaction mixture.
9. Pass the mixture down a Sephadex G50 column (5 ml pipette plugged with glass wool). Biotin labelled DNA is co-eluted with the dextran blue dye.

8.15.2 Labelling DNA Fragments by Random Priming

Procedure

1. Prepare the oligo-labelling buffer by mixing 100 μl of solution 2, 250 μl of solution 3 and 150 μl of solution 4.
2. Add sufficient SDW to a final volume of 50 μl.
3. Add 10 μl of oligo-labelling buffer to an Eppendorf tube.
4. Prepare 10 mg.ml^{-1} BSA solution and add 2 μl to the mixture.
5. Heat denature the DNA fragments (100–1000 ng) for 5 min at 100 °C and add these to the mixture.
6. Add 2 μl of biotin-7-dATP.
7. Initiate the reaction by the addition of 1 μl (2 units) of Klenow fragment to the mixture.
8. Incubate the reaction mixture for 3–5 h at 37 °C. The random priming reaction mixture may be incubated overnight at 37 °C.
9. The labelling reaction is stopped by the addition of 20 μl of stop solution.
10. Purification of the labelled products is performed as described for nick translation.

8.16 Calibration of Biotin-Labelled Probes

Labelling DNA probes by the method of nick translation requires further characterisation to ensure the appropriate size DNA fragments suitable for hybridisation are produced after heat denaturation (see Fig. 6).

Denatured DNA fragments suitable for hybridisation should be within the range 200–1000 bp. If the majority of the fragments are less or greater

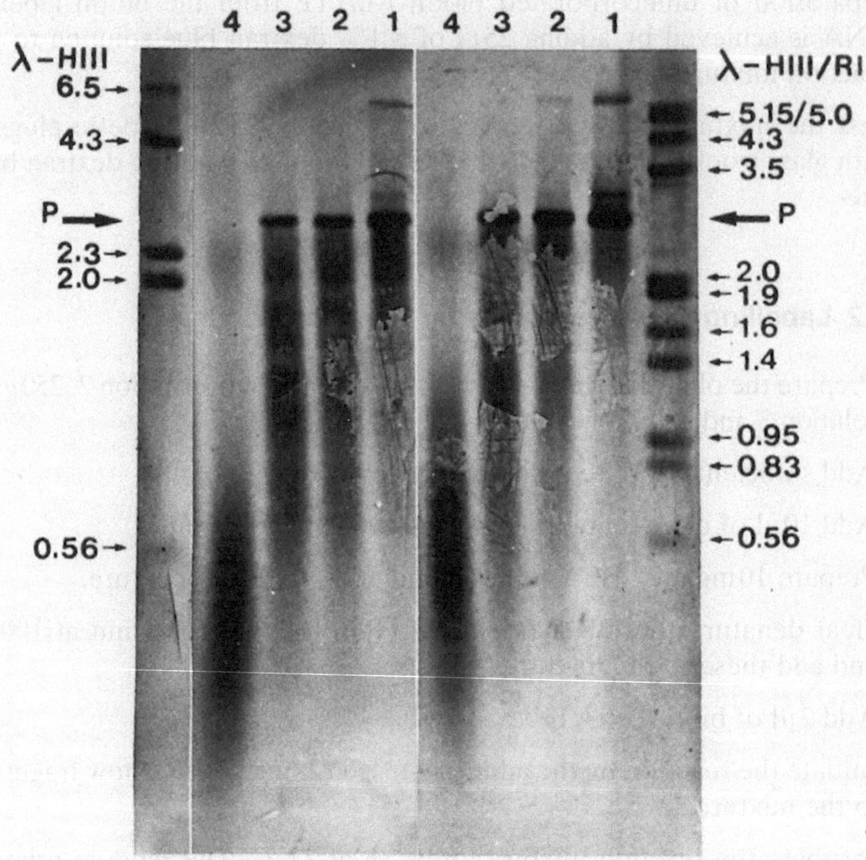

Fig. 6. An alkaline gel containing several biotin nick translation reactions that has been Southern blotted and the detection of labelled fragments with the NBT/BCIP reagents. The Lambda-HindIII (HIII) and Lambda-HindIII/Eco RI (HIII/RI) molecular weight markers are shown. Intact plasmid (*P*) containing the cloned plant DNA to be used as a hybridisation probe is shown. The numbers at the top of the gel represent the different μl of DNase I (in duplicate), that was included into the biotin nick translation reactions. A smear of DNA fragments within the range 200–1000 base pair was produced with 4 μl of DNase I

than these limits, the amount of DNase I should be adjusted to either decrease or increase the fragment sizes.

Sizing the DNA fragments is an important aspect of characterising the probe. The hybridisation of large DNA fragments may increase the background signal due to the formation of hybridisation networks (fragments binding to each other).

This procedure involves the alkaline denaturation of the probe, the resolution of the fragments by alkaline gel electrophoresis, Southern blotting and the detection of these fragments with NBT/BCIP reagents.

Equipment and materials

As for agarose gel electrophoresis (see Sect. 8.12)

Stock solutions

- 0.5 M EDTA
- 5 M NaCl
- neutralisation solution (see Sect. 8.13)
- 10 M NaOH

Reagents and solutions

- Solution containing 50 mM NaCl and 1 mM EDTA
- electrophoresis buffer (30 mM sodium hydroxide and 1 mM EDTA)
- alkaline loading buffer (50mM sodium hydroxide; 1mM EDTA; 2.5% Ficoll, type 400 and 0.025% bromocresol green)

Procedure

1. Prepare a 1% agarose gel in 50 mM NaCl and 1 mM EDTA.
2. Equilibrate the gel in electrophoresis buffer.
3. Add 2 vol ethanol to precipitate the biotin labelled probe DNA (0.5–1 μg).
4. Resuspend the pellet in 10 μl of alkaline loading buffer.
5. Load the samples and separate the DNA at 3–5 V cm^{-1} until the dye has migrated 80% of the gel.
6. Remove the gel and neutralise for 30 min prior to Southern blotting.
7. Repeat the Southern blotting procedure previously described using nitrocellulose as the membrane filter.
8. Detect the single-stranded biotin labelled DNA fragments with NBT/BCIP reagents (see Sect. 8.18).

Important note

- If the probe and reaction conditions are the same, the characterisation needs to be performed once.

8.17 Slot Blot Analysis of Biotin-Labelled DNA

Prior to hybridisation, it is important to confirm the DNA probe is biotin labelled. This check is readily performed by slot blot analysis and the colorimetric detection of biotin labelled DNA (Fig. 7).

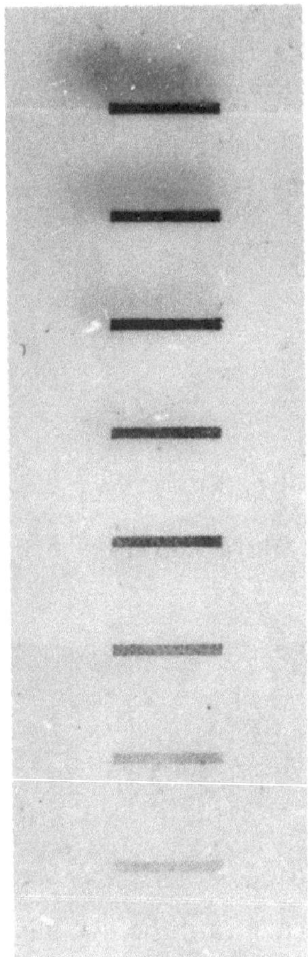

Fig. 7. Slot blot analysis of serially diluted biotin labelled DNA detected with the NBT/BCIP reagents

Equipment and materials

- micro titre tray
- Gilson pipetman

Stock solutions

- 20 × SSC
- 0.5 M EDTA
- 10 M NaOH

Reagents and solutions

- 2.5 M NaOH
- 2.5 M ammonium acetate
- 2 × SSC
- solution of 0.4 M NaOH and 1 mM EDTA

Nitrocellulose Membranes

Procedure

1. Label the DNA with biotin tagged nucleotides by the method of nick translation or random priming.
2. Transfer 1/5 volume of the reaction mixture to a microtitre tray.
3. Add an equal volume of 2.5 M NaOH, and neutralise the mixture with an equal volume of 2.5 M ammonium acetate. Neutralisation of the mixture with 2.5 M ammonium acetate is needed to avoid chemical decomposition of the nitrocellulose.
4. Serially dilute the sample with an equal volume of 20 × SSC. Serial dilution of the sample with 20 × SSC facilitates DNA binding.
5. Assemble the slot blot apparatus containing a strip fo nitrocellulose membrane pre-soaked in 2 × SSC. Ensure the slot blot apparatus is correctly assembled to avoid leaking of the samples.
6. Transfer the serially diluted samples to the slots, apply the vacuum, and wash the sample through the slots with 100 µl of 2 × SSC. Directly transfer the serially diluted samples to the membrane below the slots, this avoids loss of sample, and inconsistencies in the dilution series.
7. Remove the nitrocellulose membrane filter and bake at 80 °C for 15 min.
8. Detect the biotin labelled DNA with NBT/BCIP reagents (see Sect. 8.18).

Nylon Membranes

1. Denature an aliquot of biotin labelled DNA in a final concentration of 0.4 M NaOH and 1 mM EDTA.
2. Serially dilute the sample with an equal volume of 0.4 M NaOH and 1 mM EDTA.
3. Assemble the slot blot apparatus containing a strip of nylon membrane pre-soaked in distilled water.
4. Transfer the serially diluted samples to the slots, apply the vacuum, and wash the sample through the slots with 100 µl of 0.4 mM NaOH and 1 mM EDTA.
5. Finally apply the vacuum, and wash the sample through the slots with 100 µl of 2 × SSC.

6. Remove the nylon membrane filter and bake at 80 °C for 15 min.

7. Detect the biotin labelled DNA with NBT/BCIP reagents (see Sect. 8.18).

8.18 Colorimetric Detection of Biotin-Labelled DNA

The colorimetric detection of biotin labelled DNA exploits the high affinity interaction between biotin, streptavidin and biotin tagged molecules. Before this system can be used, biotin tagged trinucleotides must be incorporated into the DNA molecules (probe) by the described methods (nick translation or random priming). After slot blot analysis or hybridisation, the membranes are incubated with streptavidin. Streptavidin will bind to the biotin labelled DNA at one of the four available binding sites; after washing away unbound streptavidin, biotin tagged alkaline phosphatase is incubated with the membrane. Alkaline phosphatase will bind to the streptavidin to form a complex or sandwich. In the presence of 5-bromo-4-chloro-3-indolyl phosphate (BCIP), alkaline phosphatase cleaves the phosphate moiety to release protons, which reduces nitro tetrazolium blue (NBT) to produce a coloured dye precipitate.

Equipment

- bench top tilt shaker

Stock solutions

- 1 M Tris.HCl pH 7.5 and pH 9.6
- 5 M NaCl
- 1 M $MgCl_2$
- Tween 20
- NBT, 75 mg.ml^{-1} in dimethylformamide
- BCIP, 50 mg.ml^{-1} in dimethylformamide/distilled water
- 0.5 M EDTA.

Reagents and solutions

- BSA
- blocking reagent (100 mM Tris.HCl pH 7.5; 100 mM NaCl; 3 mM $MgCl_2$; 0.5% Tween 20 and 3% BSA)
- incubation reagent (100 mM Tris.HCl pH 7.5; 100 mM NaCl; 3 mM $MgCl_2$; 0.05% Tween 20 and 1% BSA)
- streptavidin (1 mg.ml^{-1})

- biotin-tagged alkaline phosphatase ($1 mg.ml^{-1}$, Sigma)
- development reagent (100mM Tris.HCl pH 9.6; 100mM NaCl; 10mM $MgCl_2$)

Procedure

1. Incubate membrane filter in 100 ml (per $100 cm^2$) of blocking reagent for 30 min.
2. Remove solution and wash membrane filter in 20 ml of incubation reagent.
3. Dilute the streptavidin solution to $1 \mu g.ml^{-1}$ with incubation reagent, and incubate membrane filter in 20 ml of this reagent for 5–10 min.
4. Remove solution and wash membrane filter three times in 50 ml of blocking reagent for 5 min.
5. Dilute biotin tagged alkaline phosphatase solution to $0.1 \mu g.ml^{-1}$ with incubation reagent, and incubate membrane filter in 20 ml of this reagent for 5–10 min.
6. Remove solution and wash membrane filter three times in 50 ml of blocking reagent.
7. Remove solution and wash membrane filter in 20 ml of development reagent.
8. Remove solution, and to 10 ml of development reagent add 44 µl of NBT and 33 µl of BCIP, and incubate in the dark until the signals are detected. After 5 to 10 min of incubation in development buffer with NBT/BCIP colorimetric signals can be observed. The background signals can also be reduced by diluting the alkaline phosphatase ($0.01 \mu g.ml^{-1}$ to reduce the background).
9. Stop the reaction by washing the membrane in 50 mM EDTA. The colorimetric reaction can be stopped before the development of the background. Where high background signals mask the slot blot signal, increasing the time in the blocking buffer can reduce this effect.

Important notes

- The serial dilution of 1/5 volume of a NT reaction (200ng) produces the following range 100, 50, 25, 6.25, 3.1, 1.5, 0.75ng etc. The efficient incorporation of biotin tagged nucleotides into the probe DNA should give a positive reaction at less than 10ng of DNA.
- Another procedure to the colorimetric detection of biotin labelled DNA can be used by substituting nylon for nitrocellulose. Slot blot analysis can

be performed with the use of nylon membranes where labelled DNA is detected by chemiluminescence. The sensitivity of detecting DNA is normally in the nano to picogram range.

8.19 Chemiluminescent Detection of the Ribosomal RNA Genes

The exposure time for the detection of DNA sequences is dependent on several factors:

- The C-value of the genome and the copy number of the target sequences in the genome.
- The efficiency of labelling probe DNA with biotin tagged nucleotides.
- Stringent hybridisation and washing conditions.
- The amount of target DNA loaded onto the membrane.
- The kinetics of light emission for the substrate 3-(29-spiroadamantane)-4-methoxy-4-(30-phosphoryloxy)phenyl-1,2-dioxetane (AMPPD).

These variables have been optimised for the detection of rDNA sequences in *S. tuberosum* (Fig. 8), and may be applicable to other species (Harding 1992). The detection of biotin labelled DNA by chemiluminescence is in accordance with the manufacturers' instructions (Tropix) with some modifications.

Equipment and materials

- bench top tilt shaker
- clear plastic wrapping film
- X-ray cassettes and X-ray film (Fuji HR-G or Kodak X-omat)
- nylon membranes and nitrocellulose (Note: Some nylon membranes and most nitro cellulose are unsuitable for the detection of DNA sequences by chemiluminescence.)

Stock solutions

Note: All solutions should be prepared from analar grade chemicals with analar or ultra-pure water, and made to final volume in the appropriate volumetric flask. All solutions should be freshly prepared.

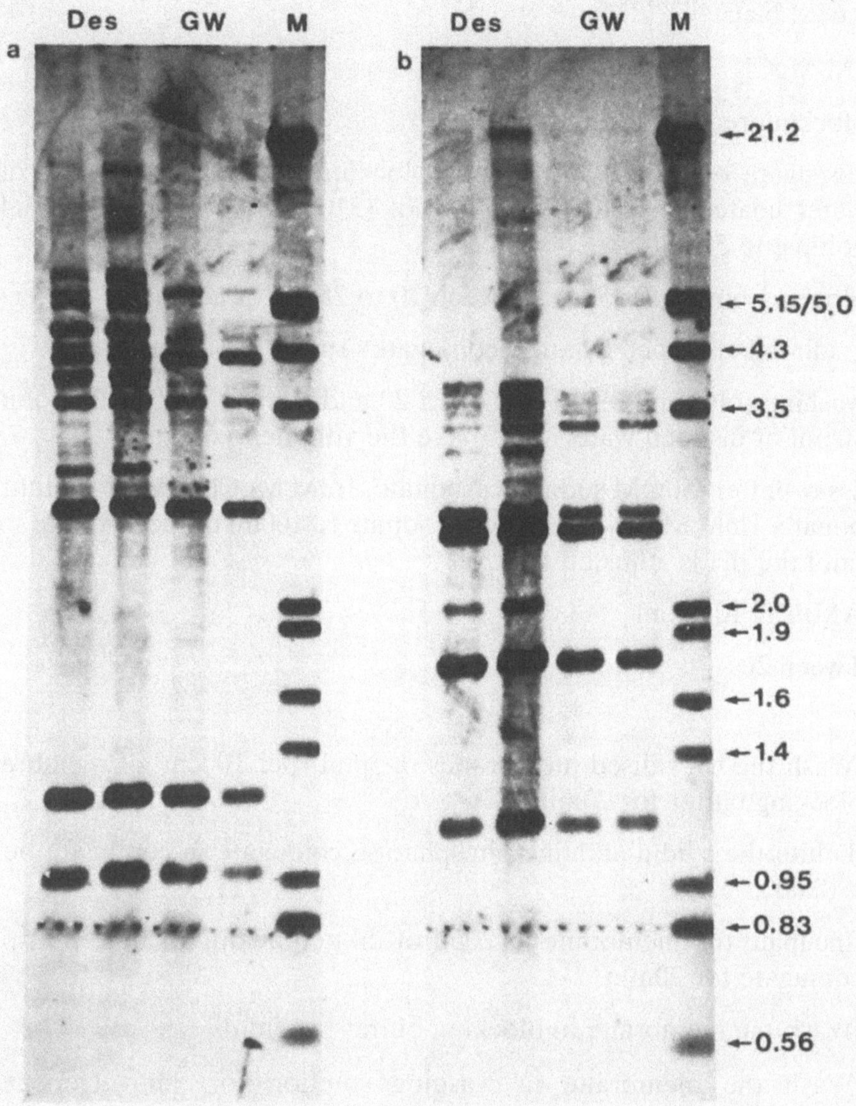

Fig. 8a,b. Detection of ribosomal RNA sequences in Bam HI (**a**) and Eco RI (**b**) duplicate samples of digested DNA derived from *S. tuberosum* cultivars Desiree (*Des*) and Golden Wonder (*GW*) by chemiluminescence on X-ray film. *M* is a Hind III/Eco RI double digestion of lambda DNA

- Phosphate buffered saline (PBS) stock solution (310) containing 0.58M disodium hydrogen phosphate; 0.17M sodium dihydrogen phosphate and 0.68M NaCl.
- 1M $MgCl_2$

Reagents and solutions

- blocking reagent (I-Light, Tropix)
- conjugate buffer (dissolve 1g of blocking reagent in 400ml of distilled water heated to 65°C. Add 50ml of (310) PBS solution and make the volume to 500ml
- blocking buffer (0.25ml of Tween 20 to 250ml of conjugate buffer)
- avidin-alkaline phosphatase conjugate (1mg.ml^{21},Sigma)
- washing solution: (1.5ml of Tween 20 and 50ml of (310) PBS solution to 400ml of distilled water, and make the volume to 500ml)
- assay buffer (50mM sodium carbonate, 1mM $MgCl_2$; 50mM sodium bicarbonate, 1mM $MgCl_2$). Mix the carbonate to 100ml of bicarbonate solution until the pH is adjusted to 9.5.
- AMPPD 10mg.ml^{-1}
- Tween 20

Procedure

1. Wash the hybridised membranes in 50ml (per 100cm^2 of membrane) of blocking buffer for 20min.
2. Dilute the avidin-alkaline phosphatase conjugate in conjugate buffer to 1:10,000.
3. Incubate the membrane in 20ml of diluted avidin-alkaline phosphatase conjugate for 20min.
4. Wash the membrane in blocking buffer for 5min.
5. Wash the membrane in washing solution for 5min (repeat three times).
6. Wash the membrane in 25ml of assay buffer for 5min (repeat twice).
7. Incubate the membrane with 10ml of assay buffer and 110Ìl of AMPPD for 5min.
8. Remove the membrane, semi-dry and seal in a clear plastic bag.
9. Chemiluminescent detection is performed in cassettes containing X-ray film without screens at room temperature.

Important notes

- The development of high background can be reduced by increasing the washing time in blocking buffer before and after incubation with the avidin-alkaline phosphatase conjugate.
- Increased background signals can be reduced by further diluting the avidin-alkaline phosphatase conjugate.
- High background signals can be reduced to a minimum by the use of ultra-pure water and chemicals in all of the reagents. Contaminating metal cations are known to be involved in the generation of oxy-free radicals and activated oxygen species. These chemical species may further interfere with the oxidative decomposition of dioxetane molecules.
- The optimal exposure time for detection is dependent on the kinetics of light emission for AMPPD (Carlson et al. 1990). Maximal emission was obtained 1 h after AMPPD incubation, where a clear fragment profile was produced after 5–10 min exposure to X-ray film. Longer exposure increased the background signal.

8.20 Analysis of Results

Restriction fragment length polymorphism (RFLP) analysis of DNA samples extracted from plants or tissues recovered from cryopreservation will detect the following DNA changes:

- The lack of a specific DNA fragment (Fig. 9).
- The presence of "new" DNA fragment.
- The variation in hybridisation signal intensity between samples (Fig. 10).

These genetic changes may be the result of mutations within the restriction enzyme target sequence, or DNA methylation of specific cytosine bases within the target sequence (Harding 1991b; Harding 1994).

Procedure

The determination of the DNA fragments molecular mass during RFLP analysis is performed by comparing the distance migrated by the DNA fragment(s) to a calibration curve. The lambda Eco RI/Hind III fragments, for example, in Fig. 6 have a known molecular weight and migration distance from the origin of the membrane. A calibration curve is constructed from plotting the molecular weights on the x axis of three cycle logarithmic graph paper, and the distance migrated at the linear scale of the y axis. The line

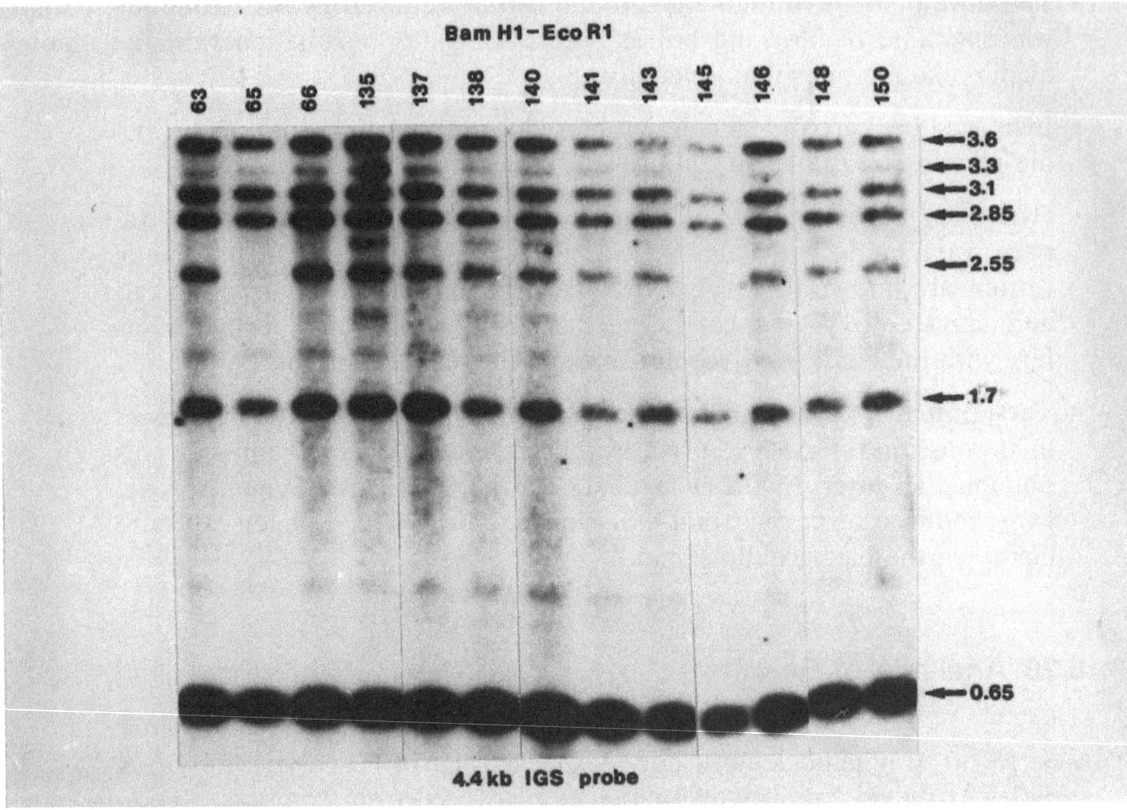

Fig. 9. Autoradiograph demonstrating RFLP differences in Bam HI-Eco RI ribosomal gene fragments in samples 65 and 145 of S. tuberosum cultivar Desiree, which lack the 2.55 kb fragment. The 4.4 kb fragment used as a hybridisation probe contains the wheat intergenic spacer (*IGS*) region with parts of the 18S and 25S ribosomal genes derived from a Bam HI-Eco RI digest of pTa71. (After Harding, 1991b)

of best fit should be linear for the fragments within the molecular weight range 0.56 to 5.0 kb, and non-linear for the higher molecular weight fragments. The molecular weight of the polymorphic DNA fragments in the samples can be obtained from the calibration curve by measuring the distance migrated.

8.21 Sensitivity of the Detection Procedures

Avidin biotin technology has been steadily developed and now has a range of applications (Wilchek and Bayer 1989). The incorporation of biotin

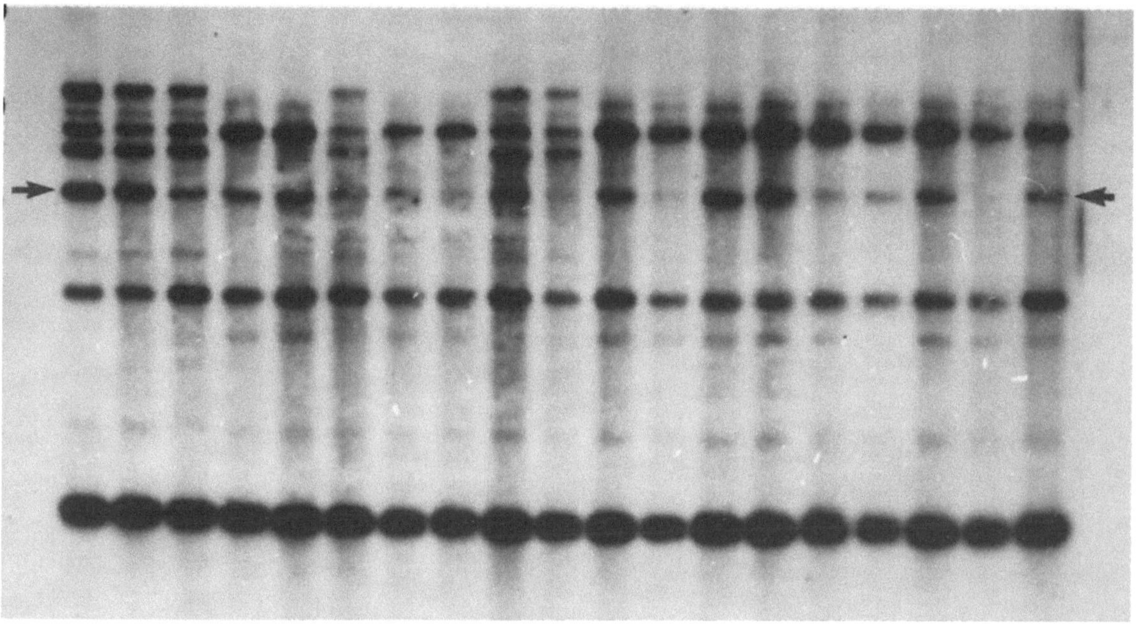

Fig. 10. Autoradiograph demonstrating variation in hybridisation signal intensity of one common DNA fragment (*arrow*) between different DNA samples of *S. tuberosum* cultivars Desiree and Golden Wonder

tagged trinucleotides into DNA and the strong affinity of avidin-enzyme conjugates has led to the colorimetric detection of DNA sequences. However, these methods have often been criticised due to low sensitivity, high background, fading hybridisation signals and lack of a suitable permanent copy of the data. The development of dioxetane substrates as substitutes for radiochemicals has also overcome the problems associated with the colorimetric detection of DNA sequences. The sensitivity of the non-radioactive detection of DNA sequences by chemiluminescence compared to conventional radio-labelled DNA is shown in Fig. 11. The results demonstrate that the sensitivity of these procedures is the same. Furthermore, the speed of detection by chemiluminescence without the associated hazards of radiation is advantageous for rapid assessments of stability in plants recovered from cryopreservation. The use of non-radioactive techniques is an expanding subject, further exemplified in the detection of single-copy DNA sequences by chemiluminescence (Kreike et al. 1990). These developments are likely to play an increasingly important role in stability studies.

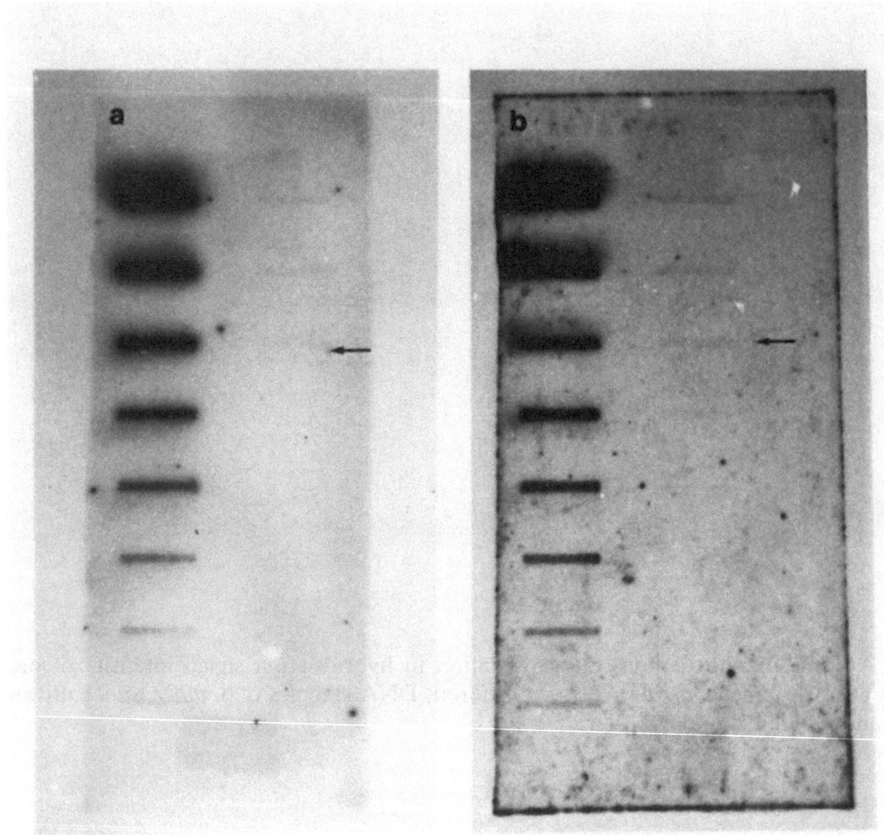

Fig. 11a,b. A photograph showing the sensitivity of radio-labelled and biotin labelled DNA. Serially diluted lambda DNA was transferred to a slot blot, radio-labelled (**a**) and biotin labelled (**b**) lambda DNA was hybridised to these separate membranes. The fragments were detected on X-ray film after 14 days by (**a**) autoradiography and 10 min by (**b**) chemiluminescence. The slots blots show the serial dilution of 1.0 μg of lambda DNA, the arrows in (**a**) and (**b**) represent the detection limit of 0.5 ng

References

Babbs CF, Pham JO, Coolbaugh RC (1989) Lethal hydroxyl radical production in Paraquat-treatd plants. Plant Physiol 90:1267–1270

Benson EE (1990) Free radical damage in stored plant germplasm. International Board for Plant Genetic Resources (Publishers) Rome

Benson EE, Hamill JD (1991) Cryopreservation and post-freeze molecular and biosynthetic stability in transformed roots of *Beta vulgaris* and *Nicotiana rustica*. Plant Cell Tissue Organ Culture 24:161–172

Benson EE, Noronha-Dutra AA (1988) Chemiluminescence in cryopreserved plant tissue cultures: the possible involvement of singlet oxygen in cryoinjury. Cryo Lett 9:120–131

Benson EE, Roubelakis-Angelakis KA (1992) Fluorescent lipid peroxidation products and antioxidant enzymes in tissue cultures of *Vitis vinifera* L. Plant Sci 84:83–90

Benson EE, Roubelakis-Angelakis KA (1994) Oxidative stress in recalcitrant tissue cultures of grapevine. Free Rad Biol and Med 16:355–364

Benson EE, Withers LA (1987) Gas chromatographic analysis of volatile hydrocarbon production by cryopreserved plant tissue cultures: a non-destructive method for assessing stability. Cryo Lett 8:35–46

Benson EE, Lynch PT, Jones J (1992a) The detection of lipid peroxidation products in cryoprotected and frozen rice cells: consequences for post-thaw survival. Plant Sci 85:107–114

Benson EE, Lynch PT, Jones J (1992b) Variation in free radical damage in rice cell suspensions with different embryogenic potentials. Planta 188:296–305

Carlson DP, Superko C, Mackey J, Gaskill ME, Hansen P (1990) Chemiluminescent detection of nucleic acid hybridization. Focus 12(1):9–12

Chevrier N, Sarhan F, Chung YS (1988) Oxidative damage and repair in *Euglena gracilis* exposed to ozone I SH groups and lipids. Plant Cell Physiol 29:321–327

Corbineau F, Engelmann F, Come D (1990) Ethylene production as an indicator of chilling injury in oil palm (Elaeis guineensis Jacq) somatic embryos. Plant Sci 71:29–34

Dillard CJ, Tappel AL (1979) Volatile hydrocarbon and carbonyl products of lipid peroxidation: a comparison of pentane, ethane, hexanal and acetone. Lipids 14:989–995

Draper HH, Hadley M (1990) Malondialdehyde determination as an index of lipid peroxidation. In: Packer L, Glazer AN (eds) Methods Enzymol 186:421–430

Eckenrode VK, Arnold J, Meagher RB (1985) Comparison of the nucleotide sequence of soybean 18S rRNA with the sequences of other small-subunit rRNAs. J Mol Evol 21:259–269

Feinberg AP, Vogelstein B (1983) A technique for radio-labelling DNA restriction endonuclease fragments to high specific activity. Anal Biochem 132:6–13

Feinberg AP, Vogelstein B (1984) ADDENDUM. A technique for radio-labelling DNA restriction endonuclease fragments to high specific activity. Anal Biochem 137:266–267

Field RJ (1981) The role of 1-aminocyclopropane-1-carboxylic acid in the control of low-temperature-induced ethylene production in leaf tissue of *Phaseolus vulgaris* L. Ann Bot 54:61–67

Field RJ (1984) The effect of low temperature on ethylene production by leaf tissue of *Phaseolus vulgaris* L. Ann Bot 47:215–223

Fraga CG, Leibovitz BE, Tappel AL (1988) Lipid peroxidation measured as thiobarbituric acid reactives substances in peroxidized lipids. Free Rad Biol Med 4:155–161

Frank H, Hintze T, Remmer H (1978) Volatile hydrocarbons in breath, an indication for peroxidative degradation of lipids. In: Kolb B (ed) Applied headspace gas chromatography, Perkin Elmer, Heyden Arrowsmith Press, London

Frankel EN (1987) Biological significance of secondary lipid peroxidation products. Free Rad Res Comm 3:213–225

Fuller BJ, Green CJ (1986) Oxidative stress in organs stored at low temperatures for transplantation. In: Rice-Evans C (ed) Free radicals cell damage and disease. Richelieu Press, London

Gerlach WL, Bedbrook JR (1979) Cloning and characterisation of ribosomal RNA genes from wheat and barley. Nucleic Acids Res 7(7):1869–1885

Green CJ, Healing G, Simpkin S, Fuller BJ, Lunec J (1986) Reduced susceptibility to lipid peroxidation in cold ishcaemic rabbit kidneys after addition of desferrioxamine, mannitol, or uric acid to the flush solution. Cryobiology 23:358–365

Gower J, Fuller BJ, Green CJ (1987) Lipid peroxidation in the cortex and medulla of rabbit kidneys subjected to cold ischaemia and the value of protective agents. Free Rad Res Comm 3:1–5

Hallinan T, Gor J, Rice-Evans CA, Stanley R, O'Reilly R (1991) Lipid peroxidation in electroporated hepatocytes occurs much more readily than does hydroxyl radical formation. Biochem J 277:767–771

Harber RM, Fuchigami LH (1986) The relationship of ethylene and ethane production to tissue damage in frozen *Rhododendron* leaf disks. J Am Soc Hortic Sci 11:434–436

Harding K (1991a) Restriction enzyme mapping of the ribosomal RNA genes in *Solanum tuberosum*: potato rDNA restriction enzyme map. Euphytica 54:245–250

Harding K (1991b) Molecular stability of the ribosomal RNA genes in *Solanum tuberosum* plants recovered from slow growth and cryopreservation. Euphytica 55:141–146

Harding K (1992) Detection of ribosomal RNA genes by chemiluminescence in *Solanum tuberosum* L.: a rapid and non-radioactive technique for the characterisation of potato germplasm. Potato Res 35:199–204

Harding K (1994) The methylation status of DNA derived from potato plants recovered from slow growth. Plant Cell, Tissue and Organ Culture 37:31–38

Heath RL, Packer l (1968) Photoperoxidation in isolated chloroplasts I Kinetics and stoichiometry of fatty acid peroxidation. Arch Biochem Biophys 125:189–198

Hebbel RP, Eaton JW, Balasingam M, Steinberg MH (1982) Spontaneous oxygen radical generation by sickle erythrocytes. Clin Invest 70:1253–1259

Jouve L, Engelmann F, Noirot M, Charrier A (1993) Evaluation of biochemical markers (sugar, proline, malondialdehyde and ethylene) for cold sensitivity in microcuttings of two coffee species. Plant Science 91:109–116

Kosugi H, Kikugawa K (1989) Potential thiobarbituric acid-reactive substances in peroxidized lipids. Free Rad Biol Med 7:205–207

Kreike CM, de Koning JRA, Krens FA (1990) Non-radioactive detection of single-copy DNA-DNA hybrids. Plant Mol Biol Rep 8(3):172–179

Levitt JA (1962) Sulfhydryl-disulfide hypothesis of frost injury and resistance in plants. J Theor Biol 3:335–391

Li XY, Chow CK (1994) An improved method for the measurement of malonialdehyde in biological samples. Lipids 29:73–75

Lunec J, Dormandy TL (1979) Fluorescent lipid-peroxidation products in synovial fluid. Clin Sci 56:53–59

Meir S, Philosoph-Hadas S, Aharon N (1992) Ethylene-increased accumulation of fluorescent lipid peroxidation products detected during senescence of parsley by a newly developed method. J Am Soc Hortic Sci 117:128–132

Murphy TM, Huerta AJ (1990) Hydrogen peroxide formation in rose cells in response to UV-C radiation. Physiol Plant 78:246–253

Rigby PWJ, Dieckmann M, Rhodes C, Berg P (1977) Labelling deoxyribonucleic acids to high specific activity in vitro by nick translation with DNA polymerase I. J Mol Biol 113:815–817

Rogers SO, Bendich AJ (1988) Extraction of DNA from plant tissues. Plant molecular biology manual, A6:1–10. Kluwer, Dordrecht

Sambrook J, Fritsch EF, Maniatis T (1989) Molecular cloning: a laboratory manual (second edition). Cold Spring Harbor Laboratory Press. Cold Spring Harbor, New York

Southern E (1975) Detection of specific sequences among DNA fragments separated by gel electrophoresis. J Mol Biol 98:503–518

Steponkus PL, Lamphear FO (1967) Refinement of the triphenyl tetrazolium chloride method of determining cold injury. Plant Physiol 42:1423–1426

Ward ACW, Benson EE, Blackhall NW, Cooper-Bland S, Powell W, Power JB, Davey MR (1993) Flow cytometric assessments of ploidy stability in cryopreseved dihaploid *Solanum tuberosum* and wild *Solanum* species. Cryo Lett 14:145–152

Widholme JM (1972) The use of fluorescein diacetate and phenosafranine for determining viability of cultured plant cells. Stain Technol 47:189
Wilchek M, Bayer EA (1989) Avidin biotin technology ten years on: has it lived up to its expectations. Trends in Biochemistry 14(10):408–412

Recommended reading

Books

Gelvin SB, Schilperoort RA, Verma DPS (eds) (1988) Plant molecular biology manual. Kluwer Dordrecht
Hames BD, Higgins SJ (eds) (1985) Nucleic acid hybridisation. IRL Press, Oxford
Kricka LJ (ed) (1992) Nonisotopic DNA probe techniques. Academic Press, New York
Packer L, Glazer AN (eds) (1990) Oxygen radicals in biological systems (part B). Methods in enzymology, vol 186 Academic Press, New York
Robyt JF, White BJ (1987) Biochemical techniques, theory and practice. Brooks/Cole, Monterey, California
Wilchek M, Bayer EA (eds) (1990) Avidin biotin technology. Methods in enzymology, vol 184. Academic Press, New York

Appendices

Abbreviations used in this chapter

AMPPD – 3-(2′-spiroadamantane)-4-methoxy-4-(3″-phosphoryloxy)phenyl-1,2-dioxetane
BCIP – 5-bromo-4-chloro-3-indolyl phosphate
bp – base pair
BSA – bovine serum albumin
$CaCl_2$ – calcium chloride
$\bullet CH_3$ – methyl radical
CH_4 – methane
C_2H_4 – ethylene
CsCl – caesium chloride
CTAB – cetyltrimethyl ammonium bromide
dATP – 2′-deoxyadenosine 5′-triphosphate
dATP-biotin – 2′-deoxyadenosine 5′-triphosphate biotin labelled
dCTP – 2′-deoxycytidine 5′-triphosphate
dGTP – 2′-deoxyguanosine 5′-triphosphate
dTTP – thymidine 5′-triphosphate
DMSO – dimethyl sulphoxide
DNA – deoxyribose nucleic acid
DTNB – 5,5′-dithiobis-(2-nitrobenzoic acid)
EDTA – ethylene diamine tetraacetic acid
FDA – fluorescein diacetate
FID – flame ionization detector
GC – gas chromatograph
GSH – glutathione
Hepes – N-2-hydroxyethylpiperazine-N′-2-ethanesulfonic acid
HCl – hydrochloric acid
kb – kilo base
KCl – potassium chloride
MDA – malondialdehyde
Mops – 3-[morpholino] propanesulfonic acid

$MgCl_2$ – magnesium chloride
•OH – hydroxyl radical
NaCl – sodium chloride
NaOH – sodium hydroxide
NBT – nitro tetrazolium blue
NT – nick translation
PVP – poly vinyl pyrolidine
RNA – ribose nucleic acid
SDS – sodium dodecyl sulphate
SDW –sterile distilled water
SH – sulfhydryl group
SSC – salt and sodium citrate solution
TAE – Tris-acetate buffer
TBA – thiobarbituric acid
TBE – Tris-borate buffer
TCA – trichloroacetic acid
TE – Tris-EDTA buffer
Tris – hydroxymethyl aminomethane
TTC – triphenyl tetrazolium chloride

List of Suppliers

For reagents and information concerning chemiluminescence:

Tropix, Inc.
47, Wiggins Avenue,
Bedford, Massachusetts 01730
USA

New Brunswick Scientific Ltd.,
North Mymms,
Hatfield,
Herts. AL9 7BR
UK

Plastic hybridisation bags:

Hybaid
111-113, Waldegrave Rd.,
Teddington,
Middlesex TW11 8LL
UK

Resin column Qiagen tubes:

Hybaid
111-113, Waldegrave Rd.,
Teddington,
Middlesex TW11 8LL
UK

QIAGEN Inc.,
11712, Moorpark Street,
P.O. Box 7401-737,
Studio City, CA 91604
USA

Malondialdehyde:

Aldrich Chemical Company, Inc.
1001 West Saint Paul Avenue,
Milwaukee,
Wisconsin 53233 USA

Gas chromatography accessories:

Pierce Eurochemie BV,
European Corporate HQ,
P.O. Box 1512,
3260 BA oud Beijerland,
Netherlands

Pierce UK Ltd
36, Clifton Rd.
Cambridge CB7 4ZR,
UK

Springer-Verlag and the Environment

We at Springer-Verlag firmly believe that an international science publisher has a special obligation to the environment, and our corporate policies consistently reflect this conviction.

We also expect our business partners – paper mills, printers, packaging manufacturers, etc. – to commit themselves to using environmentally friendly materials and production processes.

The paper in this book is made from low- or no-chlorine pulp and is acid free, in conformance with international standards for paper permanency.

MIX
Papier aus verantwortungsvollen Quellen
Paper from responsible sources
FSC® C105338

If you have any concerns about our products,
you can contact us on
ProductSafety@springernature.com

In case Publisher is established outside the EU,
the EU authorized representative is:
Springer Nature Customer Service Center GmbH
Europaplatz 3, 69115 Heidelberg, Germany

Printed by Libri Plureos GmbH
in Hamburg, Germany